AIR DRYING OF LUMBER:
A Guide to Industry Practices

By

Raymond C. Rietz

formerly research engineer
Forest Products Laboratory

and

Rufus H. Page

formerly chief, Branch of
Forest Products Utilization
Division of Cooperative Forest Management
State and Private Forestry

For the
Forest Service
U.S. Department of Agriculture

Books for Business
New York-Hong Kong

Air Drying of Lumber:
A Guide to Industrial Practices

by
Raymond C. Rietz
Rufus H. Page

for the Forest Service
U.S. Department of Agriculture

ISBN: 0-89499-216-3

Reprinted from the 1971 edition

Books for Business
New York - Hong Kong
http://www.BusinessBooksInternational.com

ABSTRACT

Describes how lumber can be dried most effectively under outdoor conditions, and illustrates air drying principles and procedures developed through field investigations and observations of industrial practices. Particular emphasis is placed on the yarding of lumber in unit packages.

Included are such subjects as why lumber is dried, advantages and limitations of the process, properties of wood in relation to drying, the drying yard layout, piling methods, causes and remedies of air-drying defects, factors affecting the cost of air drying, and protection of air-dried stock.

ACKNOWLEDGMENT

Special recognition is due Edward C. Peck, formerly wood drying specialist at the Forest Products Laboratory, for the research and related contributions that form the basis for a substantial part of this publication. The authors also appreciate the cooperation of lumber producing and using industries in providing photographs of their operations and permission to publish them.

ILLUSTRATION REQUESTS

Requests for copies of illustrations contained in this Handbook should be directed to the Forest Products Laboratory, USDA Forest Service, P.O.B. 5130, Madison, Wis. 2 53705

CONTENTS

INTRODUCTION

The tree that grows in the forest contains considerable water. Most of this water must be evaporated before the lumber obtained can be converted into consumer products. And for most wood products there is no substitute for dry lumber!

Air drying is the most economical method to remove large quantities of water from the wood, especially in the early stages of drying. Air drying may also be supplemented by other methods of drying the lumber to its final moisture content.

Drying wood by allowing nature's forces to gradually evaporate the vast supply of water is not a new idea. Over the years, however, men have learned more and more about how to help nature do a better job. But this information accumulated slowly, and many individuals have been aware of only part of the picture.

This publication assembles the information on air drying uncovered by more recent research, as well as that evolved by generations of experience.

The summation here should give plant managers, yard foremen, lumber handlers, and others in related areas a better understanding of the principles of air drying. It may also assist them in analyzing their own air-drying practices. Drying wood by any means is expensive, so progressive operators are constantly on the alert to reduce costs. Application of general principles of air drying can lead to changes in yarding methods that may result in faster drying of lumber with improved quality.

But the results of efficient air drying benefit consumers of forest products as well and are of key importance in utilization of the Nation's forest resource. In effect it helps to conserve supplies of wood, and thereby the timber resource, by reducing loss of product. In addition it helps to assure continued markets for wood products by contributing to customer satisfaction. Both are a part of the wise use of timber, long an accepted tenet of the USDA Forest Service's conservation policy.

Accent in this publication is on practicality. After touching on the variety of reasons for drying lumber and the wood properties that relate to drying, the basics of the air-drying process are spelled out. Then descriptions are given of the drying yard, methods of piling lumber for air drying, defects that occur in air drying, factors affecting cost, and protection of the air-dried lumber. Finally, a special list of recommendations near the end of the publication distills this information and serves as a convenient means of checking specific points in an individual operation.

As in any industry, specific terminology has evolved and shades of meaning show up geographically, between hardwood and softwood production, or between a plant operator and a researcher. For this reason a glossary of terms is included to delineate the way words are used in this publication.

Wood species are generally listed in the handbook by their common names. But, because different names are often used for the same species, both the common and botanical names are included in the first table.

References for additional reading are listed at the end of appropriate chapters.

CHAPTER 1
WHY DRY LUMBER?

Most of the water in the living tree must be removed before useful products can be made from the wood. The rough, green lumber sawed from the log must be dried before it is processed into most end products. Drying the lumber at this stage has a number of distinct and important advantages:

- Drying reduces weight and thereby reduces shipping and handling costs.
- The shrinkage that accompanies drying takes place before the wood is used as a product.
- As wood dries, it increases in most strength properties.
- Strength of joints made with nails and screws is greater in dry wood than in green.
- Wood must be relatively dry before it can be glued or treated with preservatives and fire-retardant chemicals.
- Drying reduces the likelihood of mold, stain, or decay.
- Drying also increases electrical resistance and thermal insulating properties and improves finishing characteristics.

Methods Used to Dry Lumber

A number of methods are employed to dry lumber, ranging from air and kiln drying to special seasoning processes. But basically all involve moving moisture from the inside of the wood to the surface, where it is evaporated into the air. Heat and air movement speed up the process. All of the major drying methods make use of either one or both of these factors.

While this handbook deals only with air drying, a brief description is included of the other major methods to clarify how they differ from air drying.

Air Drying

Lumber, arranged in layers with separating stickers, is built up into unit packages and piles outdoors so that atmospheric air can circulate through the piles and carry away moisture (fig. 1).

One modification of air drying is shed drying, where the lumber to be dried is placed in a shed with open sides (fig. 2). The roofed structure protects the lumber from rain and direct solar radiation but allows outdoor air to circulate through the stickered lumber to dry it.

Shed-Fan Air Drying

To accelerate drying, stickered unit packages of lumber are placed in an unheated shed or building that has fans on one side and is open on the other. The fans stimulate air movement through the spaces between the courses of wood (fig. 3).

Forced-Air Drying and Predrying

In these more complex processes of accelerated drying, stickered packages of lumber are placed in closed buildings that are provided with fans to recirculate heated air through the lumber piles. Both the forced-air dryers and predryers are commonly considered as low-temperature, forced-air-circulation, ventilated dry kilns. The difference is that the forced-air dryer is often built of materials that cannot be converted to a conventional dry kiln. The predryer, however, can be converted by adding more control instrumentation and by insulating the building (fig. 4).

Kiln Drying

Lumber is dried in a closed chamber by controlling the amount of heat, relative humidity, and air circulation until the wood reaches a predetermined moisture content quality (fig. 5).

M 110 053

Figure 1.—An air-drying yard arranged for good circulation of air around the piles of packaged lumber.

Special Drying Processes

To reduce drying time and drying degrade a number of processes for drying lumber have been investigated. Energy to evaporate the water in the lumber can be supplied by radiation such as in infrared heating. Radiation from the cylinder walls heats the lumber in vacuum drying. Heat for evaporation can be supplied by conduction as in boiling in oil, solvent seasoning, and press drying.

In air drying and conventional kiln drying, heat is supplied by convection. Other processes that supply heat by convection are vapor drying, impingement drying, and high-temperature drying, either with superheated steam or air and water-vapor mixtures. Special wood drying processes that utilize electrical energy for heating the wood are high-frequency dielectric heating and microwave heating.

M 137 620

Figure 2.—Drying shed for hardwood lumber.

M 135 054

Figure 3.—This shed-fan air dryer draws outdoor air through stickered unit packages of lumber that are placed crossways in the bays.

3

Figure 4.—A predryer that could be converted to a conventional forced-air-circulation, low-temperature dry kiln. Vents and fresh-air intakes are on the doors.

Choice of Methods

The factors that determine the lumber drying process used at a plant are generally related in one way or another to economics. A sawmill producing a considerable volume of a rather slow-drying wood like oak will select air drying as the cheapest process to reduce weight for shipping purposes. Softwoods, however, are often kiln dried green from the saw.

Figure 5.—Kiln drying. A steam-heated, ventilated, forced-air-condition, package-loaded lumber dry kiln being loaded with air-dried lumber.

Objectives of the Air-Drying Process

The main purpose for air-drying lumber is to evaporate as much water as possible. Free water is usually present in the cell cavities of green wood; moisture in the cell walls is called bound water. In air drying, lumber is left on stickers in the yard until it reaches a moisture content of 20 to 25 percent, indicating that the free water has been evaporated. The lumber may then be ready for further processing, depending upon its use. If it must be dried to lower moisture content levels, such as for use in furniture factories, it will be kiln dried.

When the lumber use does not require a low moisture content, air drying often is sufficient. Lumber used for outdoor furniture and other outdoor exposures, or for building structures such as barns and garages that are not heated, can usually be air dried to a low enough moisture content.

Softwoods are air dried at producing sawmills to improve machining characteristics, and to work to standard patterns and sizes only after some shrinkage has taken place. Rough-sawn hardwoods, on the other hand, are air dried at the producing sawmills primarily to reduce weight so that shipping costs are reduced.

The air-drying process, particularly for hardwoods, offers a means of carrying an inventory of various species, grades, and sizes of lumber. To meet shipping schedules during periods of the year when the sawmill cannot be operated to capacity, the yard inventory is built up when sawing conditions are favorable and the lumber is air dried while being held.

Air drying is also widely used to reduce the moisture content in wood to a level suitable for preservative treatment.

Air drying further reduces the chance that mold, stain, and decay may develop in transit, storage, or subsequent use. Blue stain and wood-destroying fungi cannot grow in wood with a moisture content of less than 20 percent. The green lumber, however, may have to be treated with a fungicide to protect it from these fungi in the early stages of the air-drying process. Drying also is a protective measure against damage from most insects that bore holes in wood.

Advantages and Limitations of Air-Drying

The real advantage of air-drying lumber over drying by other processes is its low initial cost. The cost of kiln drying dense hardwoods to the

same moisture content levels achieved in air dry
ing could be prohibitively high. However, as the
value of the wood increases, kiln drying the green
wood becomes more feasible. Species like beech,
birch, and maple are now often kiln dried green
from the saw even in the thicker sizes. Small sawmills and concentration yards that cannot afford
dry kiln installations air-dry their production before surfacing or machining to pattern.

The limitations of air drying are generally associated with the uncontrollable nature of the proc
ess. The drying rate is very slow during the cold
winter months in the northern sector of the country. At other times hot, dry winds may increase
degrade and volume losses due to severe surface
checking and end splitting. Production schedules
depend on changing climatic conditions of temperature, relative humidity, rainfall, sunshine, and
winds. Warm, humid, or sultry periods with little
air movement may encourage the growth of blue
stain and aggravate chemical brown stain. The
general principles of air drying are reasonably
well understood, however, and their application
can make this method of reducing moisture content more efficient and profitable.

For Additional Information

Brown, N. C., and Bethel, J. S.
    1958. Lumber—The stages of manufacture
        from sawmill to consumer. Ed. 2,
        John Wiley and Sons, N.Y.
Brown, W. H.
    1965. An introduction to the seasoning of tim-
        ber. Pergamon Press, N.Y.
Henderson, H. L.
    1951. The air seasoning and kiln drying of
        wood. Ed. 5, Henderson, Albany,
        N.Y.
Kozlik, C. J.
    1959. Bibliography of special seasoning
        methods. Forest Prod. Res. Center,
        Oreg. State Univ., Corvallis.
Lowery, D. P., and Krier, J. P.
    1966. Bibliography of high-temperature kiln
        drying of lumber. USDA, Forest
        Serv. Res. Pap. INT 27, Intermoun-
        tain Forest and Range Exp. Sta.,
        Ogden, Utah.
Rasmussen, E. F.
    1961. Dry kiln operator's manual. USDA
        Agr. Handb. 188.

CHAPTER 2

WOOD PROPERTIES IN RELATION TO DRYING

The structure of wood, the location and amount of moisture green wood contains, and its physical properties greatly influence its drying characteristics and reactions to air-drying conditions.

Structure of Wood

Bark, Wood, and Pith

A cross section of a tree (fig. 6) shows well-defined features from the outside to the center. The bark is divided into two layers—the outer corky, dead portion and the inner living portion. The light-colored zone next to the bark is called sapwood and darker, inner zone is called heartwood. In the center of the tree is a very small, soft core known as pith.

Hardwoods and Softwoods

Hardwoods are trees with broad leaves, and softwoods are trees with needlelike or scalelike leaves. There are exceptions to this rule, however. These terms do not apply to the hardness or softness of the woods, because some softwoods—for example southern pine—are harder than some hardwoods, like basswood or cottonwood.

The structure of hardwoods is generally more complex than that of softwoods. Figure 7 shows the pores and other cells in a hardwood cube highly magnified. Figure 8 shows a similar cube of softwood. Many hardwoods contain relatively large wood rays, while some softwoods contain resin ducts. Both rays and resin ducts are related to susceptibility to surface checking during drying.

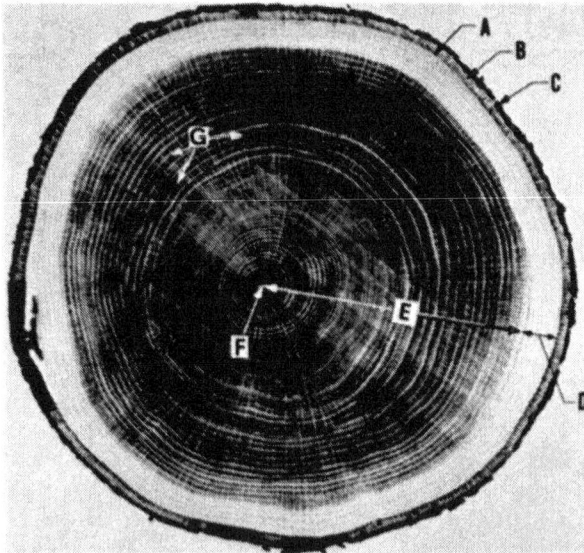

M 88620 F

Figure 6.—Cross section of an oak tree: A, Cambium layer (microscopic) is inside of inner bark and forms wood and bark cells. B, Inner bark is moist and soft, carries prepared food from leaves to all growing parts of tree. C, Outer bark or corky layer is composed of dry, dead tissue, gives general protection against external injuries. D, Sapwood is the light-colored wood beneath the bark; it carries sap from roots to leaves. E, Heartwood (inactive) is formed by a gradual change in the sapwood; it gives the tree strength. F, Pith is the soft tissue about which the first wood growth takes place in the newly formed twigs. G, Wood rays are strips of cells that extend radially within the tree and serve primarily to store and transport food.

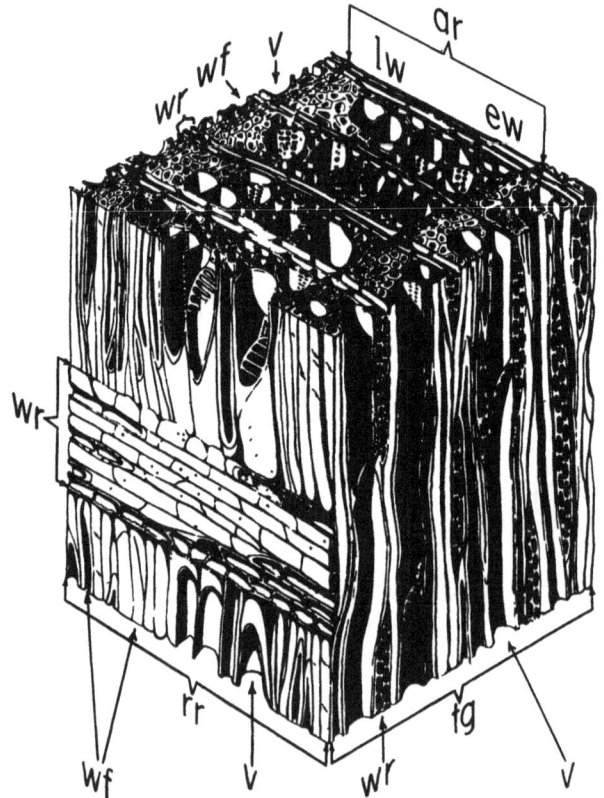

M 134 581

Figure 7.—Drawing of a section of a hardwood highly magnified. rr, radial surface; tg, tangential surface; ar, annual growth ring; ew, earlywood; lw, latewood; wr, wood ray; wf, wood fibers; v, vessels or pores.

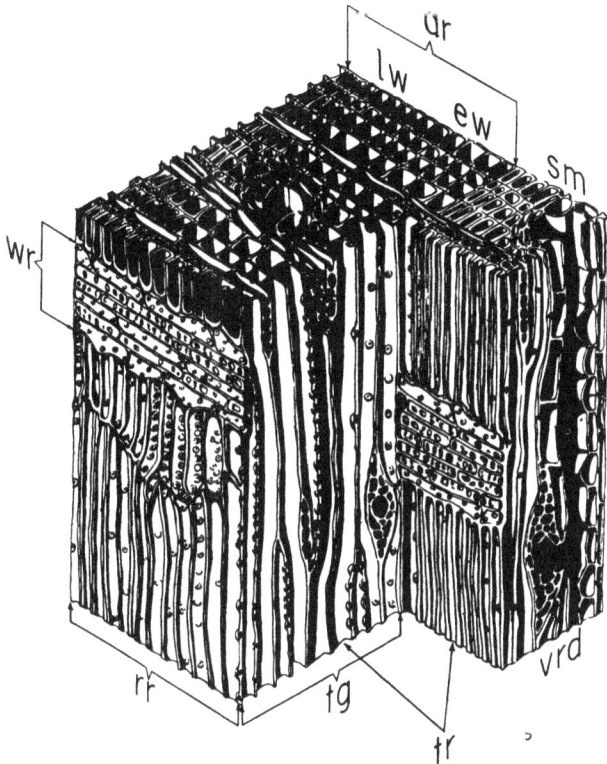

Figure 8.—Drawing of a section of a softwood highly magnified. *rr*, radial surface; *tg*, tangential surface; *ar*, annual growth ring; *ew*, earlywood; *lw*, latewood; *wr*, wood ray; *tr*, tracheid or fiber; *vrd*, vertical resin duct. The large hole near the center of the top section and the passage along the right edge (*vrd*) are vertical resin ducts.

Cellular Structure

Wood is composed of hollow, tubelike cells or fibers, usually closed at both ends. Thin spots or pit membranes are located in the walls of cells, through which the sap flows in the living tree or moisture moves during the drying of lumber.

Most of the cells lie nearly parallel to the long axis of the tree trunk, but some, the cells of the wood rays, lie on radial lines from the pith to the bark. Several layers of new cells are produced each year on the outside of the sapwood by the thin living layer called the cambium.

Earlywood and Latewood

A cross section of a tree grown in a temperate climate shows well-defined concentric layers of wood, which correspond closely to yearly increments of growth. For that reason they are commonly called annual growth rings. Earlywood— often called springwood—is formed during the early part of each growing season. The cells of the earlywood are generally larger and thinner walled than those formed later in the season; consequently earlywood is softer, weaker, and generally lighter in color than the latewood or summerwood.

Sapwood and Heartwood

In the living tree, the sapwood layer, which is next to the bark, contains many living cells that serve mainly in the transfer and storage of food. Most of its cells, however, are dead and serve only as channels for the movement of sap upward in the tree and to help support the tree. The central part of the trunk is called heartwood. All heartwood cells are dead, and their principal function is to supply strength to the trunk and to store extraneous materials.

As a tree increases in diameter by adding new layers of sapwood under the bark, the zone of heartwood enlarges at substantially the same rate. The living cells of the sapwood die and become infiltrated with gums, resins, coloring matter, and other materials. The circumference of the heartwood may be irregular, and does not necessarily follow the annual growth rings closely.

The relative amounts of sapwood and heartwood vary considerably, both between species and in trees of the same species. A tree of small diameter has more sapwood proportionately than a similar tree of larger diameter. Within a species, sapwood is thickest in the most vigorously growing trees.

Heartwood is, as a rule, less permeable to liquids than sapwood. For this reason, heartwood dries more slowly than sapwood. In resinous woods, the heartwood usually contains more resin than the sapwood.

Juvenile Wood

The wood adjacent to the pith is called juvenile wood and is characterized by indistinct growth rings. This wood shrinks more than the surrounding wood and may contribute to warping (fig. 9).

Reaction Wood

The growing tree develops wood with distinctive properties in parts of leaning or crooked trunks and in branches. This wood is called compression wood in softwoods and tension wood in hardwoods.

Compression wood especially occurs on the underside of the trunks of leaning softwood trees and on the underside of limbs. Annual growth rings in compression wood are usually wider than normal; latewood is unusually wide but does not appear as dense as normal latewood. A lack of

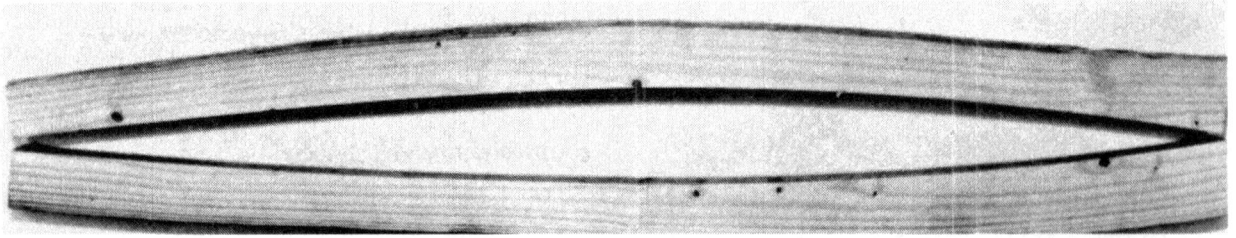

M 110 963

Figure 9.—When this board was ripped in two, each piece crooked because of longitudinal shrinkage of juvenile wood in the board center.

contrast between the earlywood and latewood gives a lifeless appearance to compression wood. It is usually yellowish or brownish in color and may also have a reddish tinge. Streaks of compression wood frequently are interspersed with normal wood. Compression wood shrinks more longitudinally than normal wood and it may cause warping or develop cross breaks during drying (figs. 10 and 11).

Tension wood occurs in hardwoods, especially on the upper side of leaning tree trunks and on the upper side of limbs. It may cause a stressed condition in the log and later contributes to splitting in sawed products. Tension wood also shrinks abnormally, and thus may contribute to warping during drying. When lumber is machined, zones of tension wood are indicated by torn grain (fig. 12).

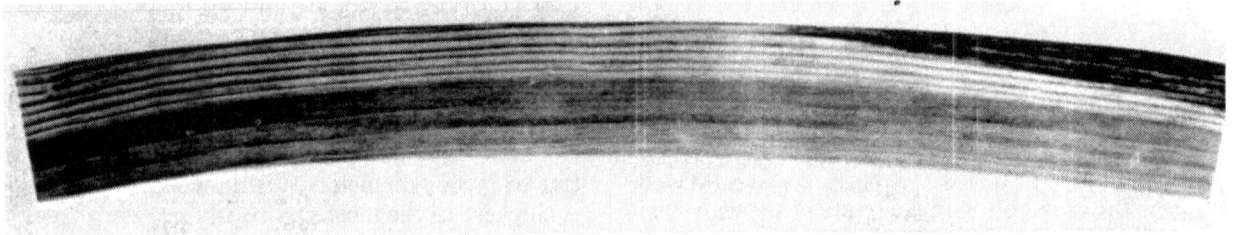

M 41251 F

Figure 10.—Crook resulted when the band of compression wood (darker strip running length of piece) shrank more longitudinally than the lighter colored normal wood.

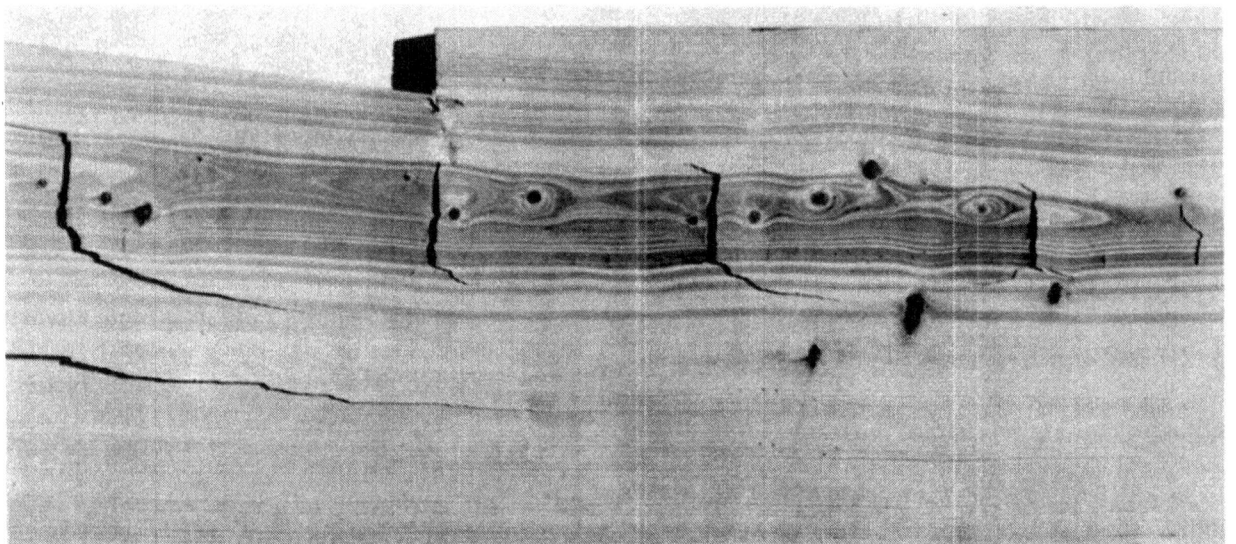

M 107 720

Figure 11.—Board with dark band of compression wood with cross breaks caused by longitudinal shrinkage.

8

M 81915 F

Figure 12.—Zones of tension wood are indicated by grain torn during machining.

Structural Irregularities

The length of the wood cells is usually parallel to the length of the tree trunk. Sometimes the length of the wood cells forms an angle with the length of the tree trunk. If this orientation of the fibers continues around the circumference and upward in the trunk, a spiral grain results. If a log is sawed parallel to the pith rather than to the bark, the boards will have a diagonal grain.

Both spiral and diagonal grain are called cross grain in boards or other sawed products. The principal drying defect resulting from cross grain is warping. Cross grain also causes mechanical weakness.

Knots

A knot is revealed when lumber is cut from the portion of a tree containing an embedded branch. Normally, a knot starts at the pith. Knots are generally objectionable because the distortion and the discontinuity of the grain around knots weakens the wood. Furthermore they cause irregular shrinkage and warping. When lumber dries, the knots and the wood adjacent to them tend to check. The knots may also become loose because of their change in size.

Moisture Content of Green Wood

The moisture in freshly cut lumber is often called sap. Sap is composed primarily of water, with varying amounts of other materials in solution. The moisture in the wood exists in two forms: As free water in the cell cavities and as bound moisture held within the cell walls. Sapwood usually contains more moisture than heartwood, particularly in softwoods. Table 1 gives moisture content values for green wood of a number of species.

The moisture content of wood may vary at different heights in the tree. Species and growing conditions are involved. Butt logs of sugar pine, western larch, redwood, and western redcedar often sink in water, although the upper logs from the same trees float. In addition to having a higher moisture content, these "sinker" logs may also have a higher specific gravity or more wood substance per unit volume.

Methods of Determining Moisture Content

The performance of wood is influenced by the amount of moisture it contains; for many uses wood serves best at specific levels of moisture content. Therefore, knowing the moisture content is an essential part of determining wood's readiness for use. The amount of moisture in wood is expressed as a percentage of the weight of the dry wood substance. Moisture content of wood may be determined by the ovendrying method, by a distillation method, or by the use of electric moisture meters.

Ovendrying Method

The ovendrying method might be called the standard way of determining moisture content in wood. A cross section is cut from a board and then completely dried in a heated oven. Generally, the

9

TABLE 1.—*Average moisture content of green wood, by species*

Species (Common name)	Botanical name	Moisture content [1] Heartwood	Sapwood	Mixed heartwood and sapwood
SOFTWOODS		*Pct*	*Pct*	*Pct*
Baldcypress	Taxodium distichum.	121	171	
Cedar:				
Alaska	Chamaecyparis nootkatensis.	32	166	
Atlantic white	Chamaecyparis thyoides.			35
Eastern redcedar.	Juniperus virginiana.	33		
Incense	Libocedrus decurrens.	40	213	
Northern white.	Thuja occidentalis.			55
Port-Orford	Chamaecyparis lawsoniana.	50	98	
Western redcedar.	Thuja plicata	58	249	
Douglas-fir:[2]				
Coast	Pseudotsuga menziesii.	37	115	
Interior north	P. menziesii	37	130	
Interior south	P. menziesii	30	130	
Interior west	P. menziesii	30	140	
Fir:				
Balsam	Abies balsamea	120	140	
California red	A. magnifica			108
Grand	A. grandis	91	136	
Noble	A. procera	34	115	
Pacific silver	A. amabilis	55	164	
Subalpine	A. lasiocarpa			47
White	A. concolor	98	160	
Hemlock:				
Eastern	Tsuga canadensis	97	119	
Western	T. heterophylla	85	170	
Larch, western	Larix occidentalis.	54	119	
Pine:				
Eastern white	Pinus strobus			68
Jack	P. banksiana			70
Lodgepole	P. contorta	41	120	
Ponderosa	P. ponderosa	40	148	
Red	P. resinosa	32	134	
Pine: Southern:				
Loblolly	P. taeda	33	110	
Longleaf	P. palustris	31	106	
Shortleaf	P. echinata	32	105	
Slash	P. elliottii	30	100	
Sugar	P. lambertiana	98	219	
Western white	P. monticola	62	148	
Redwood:				
Old-growth	Sequoia sempervirens.	86	210	
Young-growth	S. sempervirens	100	200	
Spruce:				
Engelmann	Picea engelmannii	51	173	
Red	P. rubens			55
Sitka	P. sitchensis	41	142	
White	P. glauca			55

Species (Common name)	Botanical name	Moisture content [1] Heartwood	Sapwood	Mixed heartwood and sapwood
HARDWOODS		*Pct*	*Pct.*	*Pct*
Alder, red	Alnus rubra		97	
Ash:				
Black	Fraxinus nigra	95		
Green	F. pennsylvanica		58	
White	F. americana	46	44	
Aspen:				
Bigtooth	Populus grandidentata.	95	113	
Quaking	P. tremuloides	95	113	
Basswood, American	Tilia americana	81	133	
Beech, American	Fagus grandifolia	55	72	
Birch:				
Paper	Betula papyrifera	89	72	
Sweet	B. lenta	75	70	
Yellow	B. alleghaniensis	74	72	
Butternut	Juglans cinerea			104
Cherry, black	Prunus serotina	58		65
Cottonwood:				
Black	Populus trichocarpa.	162	146	
Eastern	P. deltoides	160	145	
Elm:				
American	Ulmus americana.	95	92	
Rock	U. thomasii	44	57	
Hackberry	Celtis occidentalis.	61	65	
Hickory	Carya spp	71	51	
Magnolia, southern.	Magnolia grandiflora.	80	104	
Maple:				
Bigleaf	Acer macrophyllum.	77	138	
Red	A. rubrum			70
Silver	A. saccharinum	58	97	
Sugar	A. saccharum	65	72	
Oak:				
Northern red	Quercus rubra	80	69	
Northern white.	Q. alba	64	78	
Southern red	Q. falcata	83	75	
Southern white (chestnut).	Q. prinus	72		
Pecan	Carya illinoensis	71	62	
Sweetgum	Liquidambar styraciflua.	79	137	
Sycamore, American.	Platanus occidentalis.	114	130	
Tanoak	Lithocarpus densiflorus.			89
Tupelo:				
Black	Nyssa sylvatica	87	115	
Water	N. aquatica	150	116	
Walnut, black	Juglans nigra	90	73	
Willow, black	Salix nigra			139
Yellow-poplar	Liriodendron tulipifera.	83	106	

[1] Based on ovendry weight.
[2] Coast Douglas-fir is defined as Douglas-fir growing in the States of Oregon and Washington west of the summit of the Cascade Mountains. Interior West includes the State of California and all counties in Oregon and Washington east of but adjacent to the Cascade summit. Interior North includes the remainder of Oregon and Washington and the States of Idaho, Montana, and Wyoming. Interior South is made up of Utah, Colorado, Arizona, and New Mexico.

section remains in the oven for at least 12 hours and sometimes longer. The method consists of the five following steps:

(1) Cut a cross section about 1 inch thick, along the grain, from a board, usually some distance from the end.

(2) Immediately after sawing, remove all loose splinters and weigh the section.

(3) Put the section in an oven maintained at $103° \pm 2°$ C. ($217.4° \pm 3.6°$ F.) and dry until a constant weight is attained.

(4) Weigh the dry section to obtain the ovendry weight.

(5) Subtract the ovendry weight from the initial weight, and divide the difference by the overdry weight, multiplying the result by 100 to obtain the percentage of moisture in the section.

$$\text{Moisture content in percent} = \frac{\text{Initial weight} - \text{Ovendry weight}}{\text{Ovendry weight}} \times 100.$$

A short-cut formula convenient to use with slide rules or calculators is:

$$\text{Moisture content in percent} = \frac{\text{Initial weight}}{\text{Ovendry weight}} - 1 \times 100.$$

Several types of balances are used in weighing specimens for moisture content determinations. One of the most commonly used is the triple-beam balance. Where a considerable number of specimens are weighed in and out of the drying oven, a direct-reading automatic balance is very convenient (fig. 13). Self-calculating balances are also available (fig. 14). The moisture content calculation is carried out on the balance by following the prescribed sequence of operations supplied by the manufacturer.

The ovens used for drying the moisture sections should be large enough to accommodate a number of specimens with space between them. Ovens should also be well ventilated to allow the evaporated moisture to escape. The temperature of the oven must be controlled with a reliable thermostat. Excessive temperatures will char the specimens, introducing errors in the moisture analysis. Electric ovens are generally used but, when the volume of specimens to be ovendried daily is relatively small, a natural-circulation electric oven may be used. When large numbers of specimens are being ovendried continually, electric ovens with fans to circulate the heated air are recommended.

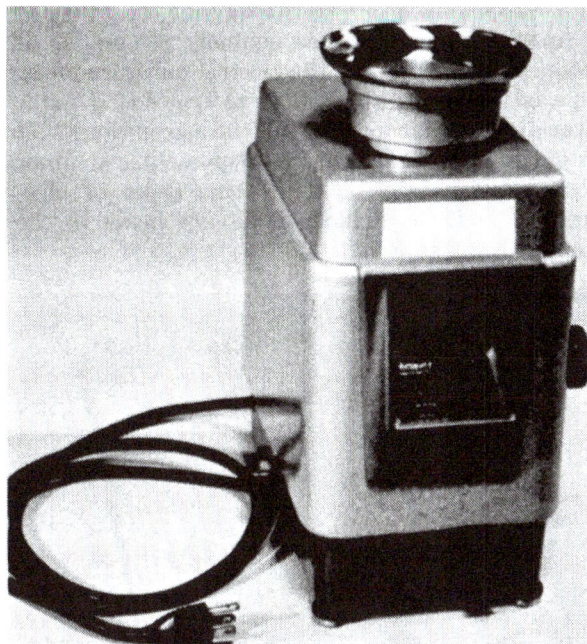

M 133 688

Figure 13.—Direct-reading automatic balance.

Distillation Method

The distillation method is used when the lumber contains volatiles such as pitch. The procedure for selecting the specimen for moisture analysis is the same as for ovendrying. The method generally consists of boiling the wood sample in toluene to dissolve the pitch deposits. Water is distilled from the wood at the same time and is collected in a condenser. By measuring the weight of the water collected, and the weight of the dry wood sample, the moisture content of the extractive-free wood can be calculated.

Electrical Methods

Electric moisture meters permit the determination of moisture content without cutting or seriously marring the board. Such meters are rapid

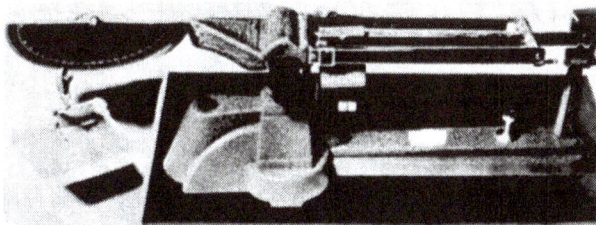

M 90343 F

Figure 14.—Self-calculating moisture content balance. The triple-beam balance is provided with a special scale on the specimen pan that is used to calculate the moisture section after the section is ovendried.

and reasonably accurate throughout the range of 7 to 25 percent moisture content. A number of tests can be made with the electric moisture meter on a board or a lot of lumber to arrive at a better overall average, canceling out the inaccuracies that might exist in individual measurements. A minor consideration is that, with some types of electrodes, small needlepoint holes are made in the lumber being tested.

Electric moisture meters are being extensively used, particularly the portable or hand meters. Three types, each based on a different fundamental relationship, have been developed: (1) The resistance type, which uses the relationship between moisture content and direct-current resistance (fig. 15); (2) the radiofrequency power-loss type, which uses the relationship between moisture content and the dielectric loss factor of the wood (fig.

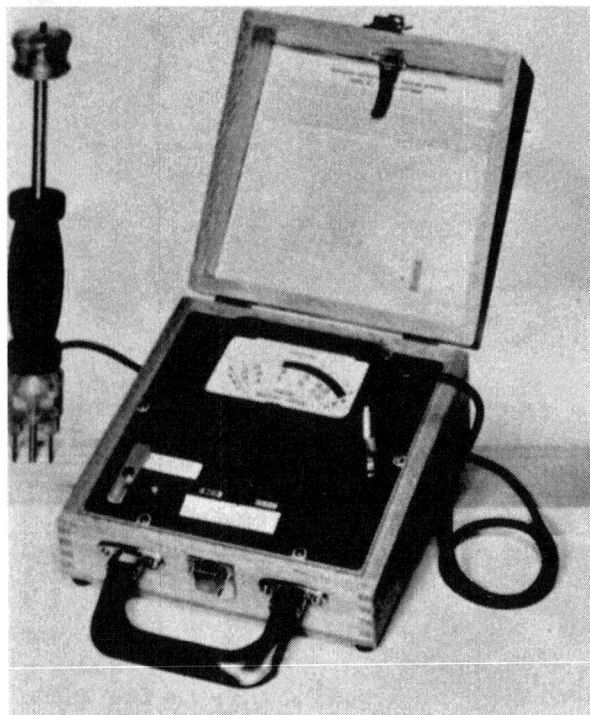

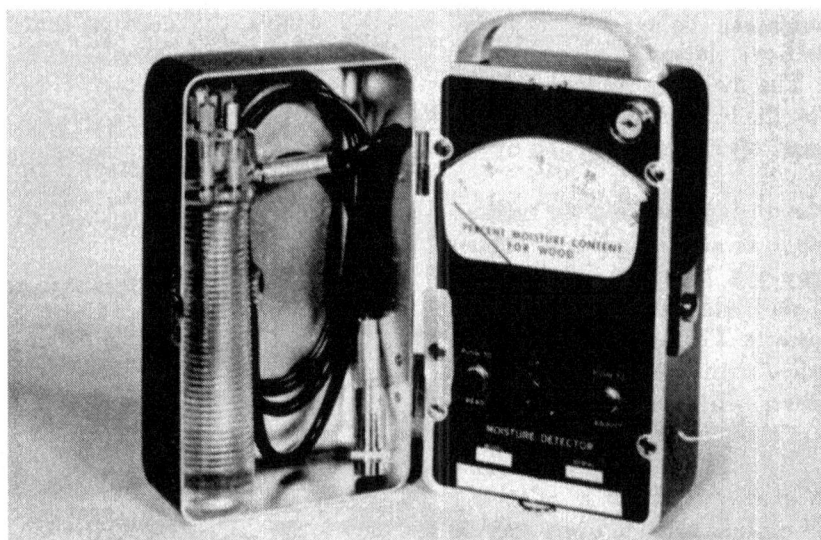

Figure 15.—Different sizes of resistance-type moisture meters.

M 134 790

12

16); and (3) the capacitance type, which uses the relationship between moisture content and the dielectric constant of the wood (fig. 17).

Portable, battery-operated, resistance-type moisture meters are wide-range ohmmeters. Most models have a direct-reading meter, calibrated in percent for one species; the manufacturer provides correction tables for other species. The manufacturer· also provides a temperature-correction chart or table for correcting the meter reading when tests are made on wood warmer than 90° F. or cooler than 70° F. If a temperature-correction chart or table is not provided by the instrument manufacturer, use the chart in figure 18.

To use figure 18, first find meter reading on vertical left margin, then follow horizontally to vertical line that corresponds to the temperature of the wood; the true moisture content can be approximated from the curved line. Example: If meter indicated 18 percent on wood at 120° F., true moisture content would be 14 percent. This chart is based on a calibration temperature of 70° F.

Resistance-type meters are generally supplied with pin-type electrodes that are driven into the wood being tested. The most common pin type electrodes use four phonograph needles that extend about five-sixteenth of an inch beyond their mounting chucks. Long, two-pin electrodes that may be insulated except for the tip are also available for use on lumber thicker than 1 5/8 inches. Valid estimates of the average moisture content of the drying board are obtained by driving the pins deep enough in the lumber so that the tip reaches one-fifth to one-fourth of the thickness of the board. If the surface of the lumber has been wet by rain, the meter indications are likely to be much higher than the actual average moisture content unless insulated shank pins are used.

Radiofrequency power-loss and capacitance-type hand meters use surface-contact-type electrodes. The electrode is an integral part of the instrument. The electric field radiating from the electrode penetrates about three-fourths of an inch into the wood so that lumber thicknesses to about 1½ inches may be tested. The moisture content of the surface layers of the lumber, however, has a predominant effect on the meter readings, simply because the electric field is stronger near the surface in contact with the electrode.

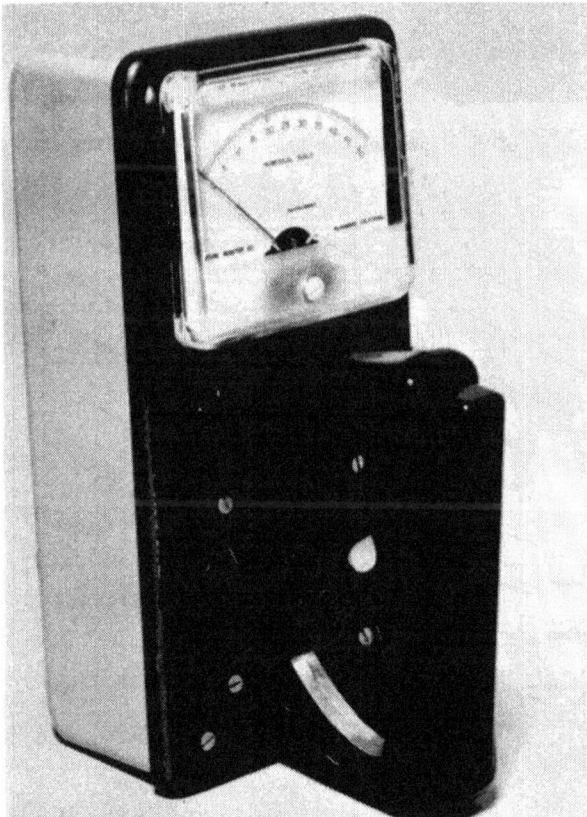

Figure 16.—A radiofrequency power-loss type of electric moisture meter.   M 133 689

13

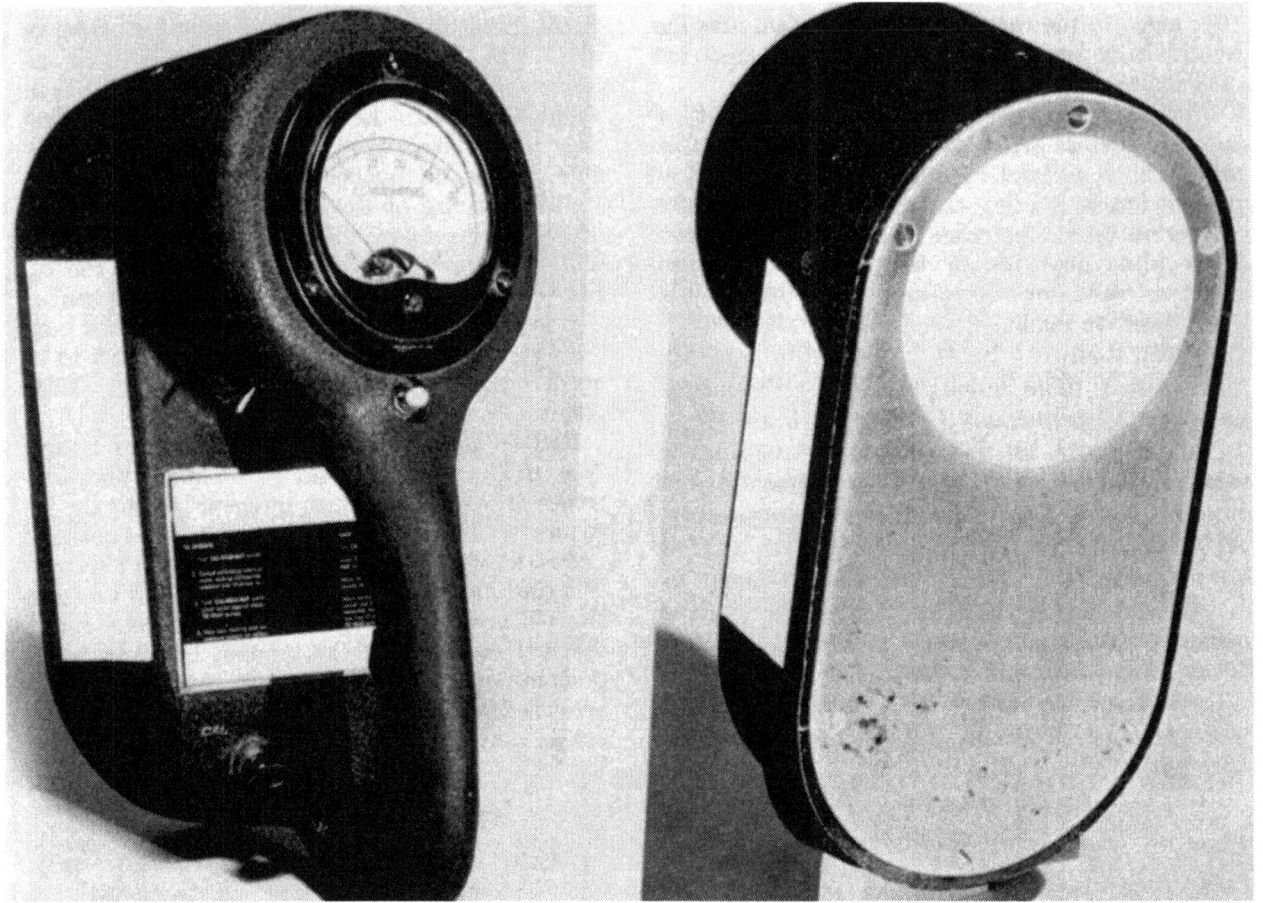

Figure 17.—Hand moisture meter for wood of the capacitance type.

M 133 691

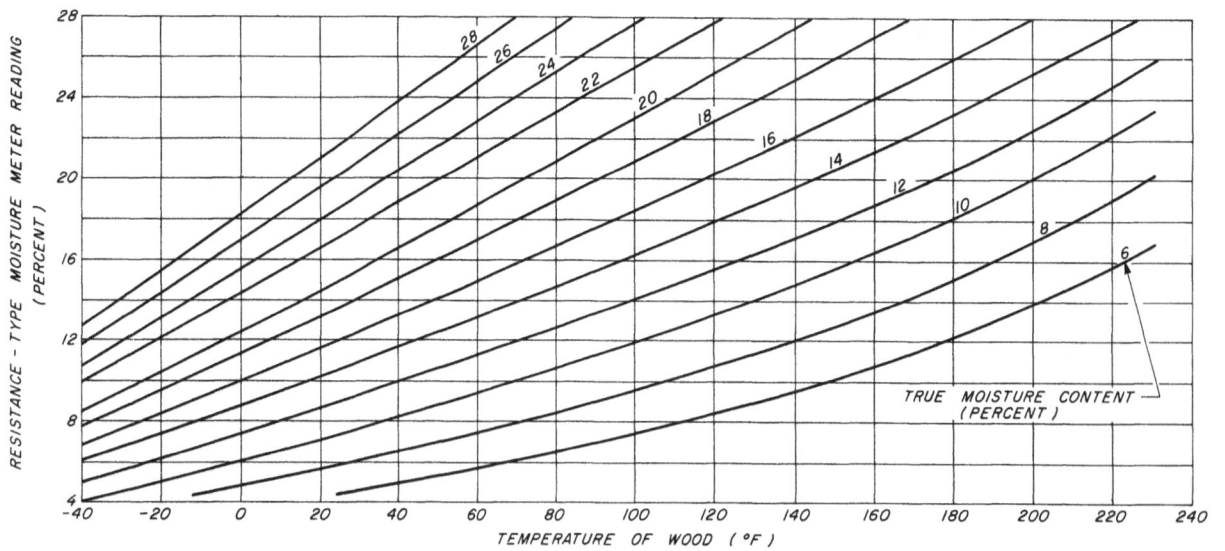

ZM 76476 F

Figure 18.—Temperature-correction chart for use with resistance-type electric moisture meters, based on combined data from several investigators.

14

Temperature-correction charts or tables are not provided by the manufacturers of power-loss and capacitance-type meters. They do, however, provide species-correction tables for converting the meter-scale reading of the instrument to moisture content for individual species.

Movement of Moisture in Wood

Water in lumber moves from zones of high moisture content to zones of lower moisture content in an effort to reach a moisture equilibrium throughout the board. This commonly means that, in drying, moisture from the interior of the wet board moves to the drier surface zones. In air drying, outdoor air must be encouraged to circulate over the green lumber to dry the surfaces and draw moisture from the interior of the board.

Passageways for Moisture Movement

Moisture moves through several kinds of passageways in the wood. The principal ones are the cavities in the cells, the pit chambers and pit-membrane openings in the cell walls, and the transitory cell-wall passageways. Movement of moisture in these passageways occurs not only lengthwise in the cells but also sidewise from cell to cell toward the drier surfaces of the wood.

Forces That Move Moisture in Wood

When wood dries, several moisture-driving forces may be operating to reduce its moisture content. These forces, which may be acting at the same time, are:

(1) Capillary action that causes the free water to move through the cell cavities, pit chambers, and pit-membrane openings.

(2) Vapor-pressure or relative-humidity differences that cause moisture in the vapor state to flow through cell cavities, pit chambers, pit-membrane openings, and intercellular spaces.

(3) Moisture content differences that cause movement of moisture through transitory passageways within the cell walls.

When green wood starts to dry, evaporation of water from openings in the surface cells creates a pull on the free water in the cell cavity and in adjacent cells. Free water moves from one tubular cell to another toward the wood surface by capillary action. When the free water has evaporated, the moisture remaining in the wood is in the form of vapor in the cell cavities or bound water in the cell walls.

The movement of water vapor through void spaces in wood depends on how much water vapor is contained in the air in the voids or in the air surrounding the wood.

If the air surrounding the wood has a low relative humidity, water vapor will move from the wet wood to the air. Thus, rapid drying depends upon the surface moisture content of the wood being dried, and whether a difference in moisture content can be developed between the surface and the interior of a board.

The higher the temperature to which the drying wood is subjected, the faster will be the combined vapor and moisture movement. This is why the air-drying rate is faster in the summer months than in wintertime.

Equilibrium Moisture Content of Wood

Wood is a hygroscopic material; it gives off or takes on moisture until it is in balance with the relative humidity and temperature of the air surrounding it. When wood has attained such a balance, it is said to have reached its equilibrium moisture content, often abbreviated to EMC. The EMC of wood can be predicted by knowing the temperature and relative humidity conditions of the air circulating around the drying lumber.

The temperature of the air in a lumber drying yard is determined with an ordinary dry-bulb thermometer. To determine the relative humidity of the air, a wet-bulb thermometer must also be provided. If the two thermometers are mounted on a single base, the instrument is called a hygrometer (fig. 19). The wet-bulb thermometer has the bulb covered with a clean, soft cloth wick which dips into a reservoir of pure, clean water. Because of the evaporation from the wet surface of the wick, the wet-bulb thermometer will give a lower reading than the dry-bulb thermometer. This difference in readings, called the wet-bulb depression, is a measure of the relative humidity of the air. To obtain a true reading of the wet-bulb thermometer, the hygrometer must be placed in a strong current of air.

Charts and tables are used to convert the hygrometer readings to a relative humidity at the dry-bulb temperature. The EMC can be determined from various charts and tables if the dry-bulb temperature and relative humidity are known. In figure 20, however, the calculation of relative humidity is eliminated as the EMC is determined directly from the dry- and wet-bulb readings of the hygrometer.

15

ber, the free water in the cells of the surface layer will leave the cell first; no bound water will leave until essentially all the free water has evaporated. The moisture condition when all the free water in the cell has evaporated, but the cell wall remains saturated, is called the fiber saturation point. For most practical purposes, this condition exists at a cell moisture content of about 30 percent. Note that the definition of the fiber saturation point applies to the cell and not to the moisture content of a board.

The fiber saturation point is critical in the drying of wood. Comparatively large changes result in the physical and mechanical properties of the wood with changes of moisture content around this point. Because the moisture removed below this point comes from the cell wall, the fiber saturation point is the moisture content level at which shrinkage starts. Furthermore, more energy is required to evaporate the bound water because the attraction between the wood and water must be overcome.

Shrinkage of Wood

When the cells in the surface layers of a board dry below the fiber saturation point, or about 30 percent, the cell walls shrink. Shrinkage of cells in the surface region of the board can be sufficient to squeeze the core and cause a slight overall shrinkage of the board. Typical moisture content-shrinkage curves are shown in figure 22. For most practical purposes the shrinkage of wood is considered as being directly proportional to the amount of moisture lost below 30 percent.

Shrinkage varies with the species and with the orientation of the fibers in the piece. Normally it is expressed as a percentage of the green dimension. The reduction in size parallel to the growth ring, or circumferentially, is called tangential shrinkage. The reduction in size parallel to the wood rays, or radially, is called radial shrinkage. Tangential shrinkage is about twice as great as radial shrinkage in most species. This explains the responses to shrinkage of several wood shapes shown in figure 23.

A plainsawed board (fig. 24) shrinks tangentially in width and radially in thickness. A quartersawed board shrinks radially in width, tangentially in thickness. Table 2 gives tangential and radial shrinkage values for the wood of many species. The longitudinal shrinkage of wood is generally slight, 0.1 to 0.2 percent of the green dimension. If reaction wood and juvenile wood are in the board, the longitudinal shrinkage may be appreciably increased.

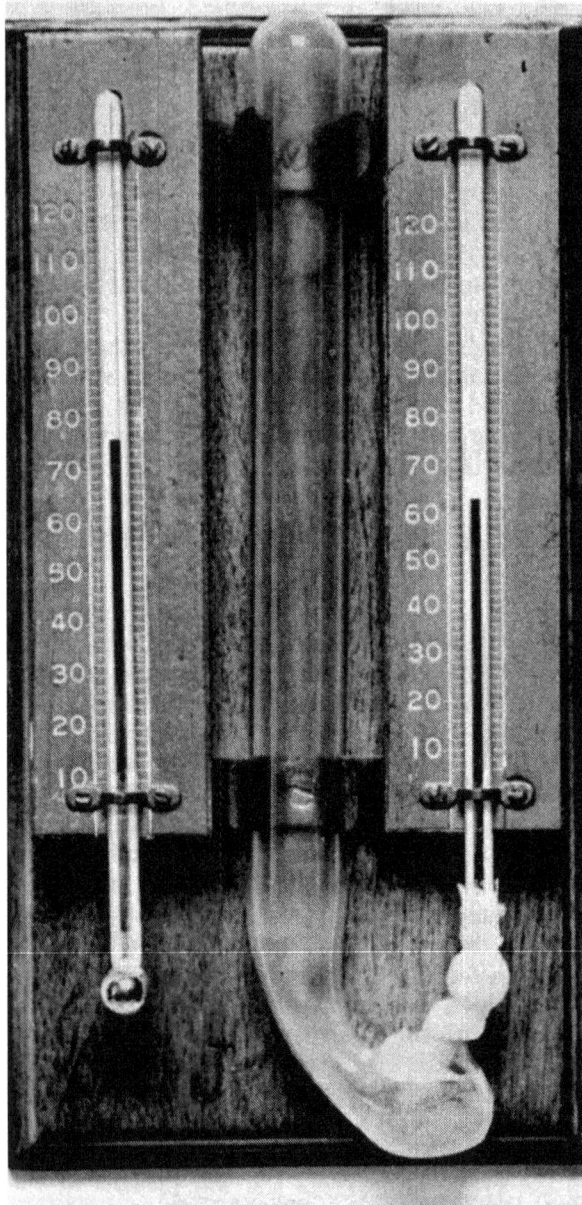

M 137 003

Figure 19.—Dry- and wet-bulb hygrometer.

The relative humidity of the air at the dry-bulb temperature can be estimated from figure 21. The EMC of the wood can also be estimated from figure 21.

Fiber Saturation Point of Wood

As previously stated, the moisture in the cell cavity is called free water and that held in the cell walls is called bound water. In drying green lum-

TABLE 2.—*Total shrinkage values of domestic woods*[1]

Species	Shrinkage to 0 percent moisture content		Species	Shrinkage to 0 percent moisture content	
	Radial	Tangential		Radial	Tangential
SOFTWOODS			**HARDWOODS**		
	Pct.	*Pct.*		*Pct.*	*Pct.*
Baldcypress	3.8	6.2	Alder, red	4.4	7.3
Cedar:			Ash:		
Alaska	2.8	6.0	Black	5.0	7.8
Atlantic-white	2.9	5.4	Green	4.6	7.1
Eastern redcedar	3.1	4.7	White	4.9	7.8
Incense	3.3	5.2	Aspen:		
Northern white	2.2	4.9	Bigtooth	3.3	7.9
Port-Orford	4.6	6.9	Quaking	3.5	6.7
Western redcedar	2.4	5.0	Basswood, American	6.6	9.3
Douglas-fir:			Beech, American	5.5	11.9
Coast	4.8	7.6	Birch:		
Interior north	3.8	6.9	Paper	6.3	8.6
Interior south			Sweet	6.5	9.0
Interior west	4.8	7.5	Yellow	7.3	9.5
Fir:			Butternut	3.4	6.4
Balsam	2.9	6.9	Cherry, black	3.7	7.1
California red	4.5	7.9	Cottonwood:		
Grand	3.4	7.5	Black	3.6	8.6
Noble	4.3	8.3	Eastern	3.9	9.2
Pacific silver	4.4	9.2	Elm:		
Subalpine	2.6	7.4	American	4.2	9.5
White	3.3	7.0	Rock	4.8	8.1
Hemlock:			Hackberry	4.8	8.9
Eastern	3.0	6.8	Hickory	7.4	11.4
Western	4.2	7.8	Magnolia, southern	5.4	6.6
Larch, western	4.5	9.1	Maple:		
Pine:			Bigleaf	3.7	7.1
Eastern pine	2.1	6.1	Red	4.0	8.2
Jack	3.7	6.6	Silver	3.0	7.2
Lodgepole	4.3	6.7	Sugar	4.8	9.9
Ponderosa	3.9	6.2	Oak:		
Pine:			Northern red	4.0	8.6
Red	3.8	7.2	Northern white	5.6	10.5
Southern:			Southern red	4.7	11.3
Loblolly	4.8	7.4	Southern white (chestnut)	5.3	10.8
Longleaf	5.1	7.5	Pecan	4.9	8.9
Shortleaf	4.6	7.7	Sweetgum	5.3	10.2
Slash	5.4	7.6	Sycamore, American	5.0	8.4
Sugar	2.9	5.6	Tanoak	4.9	11.7
Western white	4.1	7.4	Tupelo:		
Redwood:			Black	5.1	8.7
Old-growth	2.6	4.4	Water	4.2	7.6
Young-growth	2.2	4.9	Walnut, black	5.5	7.8
Spruce:			Willow, black	3.3	8.7
Engelmann	3.8	7.1	Yellow-poplar	4.6	8.2
Red	3.8	7.8			
Sitka	4.3	7.5			
White	4.7	8.2			

[1] Expressed as a percentage of the green dimension.

The values in table 2 can be converted into dimensional changes, in inches, by using the following formula:

$$S = \frac{(M_I - M_F)D}{\left(\dfrac{30}{S_T \text{ or } S_R} - 30\right) + M_I}$$

where S is shrinkage or swelling in inches; M_I is initial moisture content in percent; M_F is final moisture content in percent; D is dimension at initial moisture content in inches; 30 is fiber saturation point in percent; S_T is total tangential shrinkage, in percent, divided by 100; and S_R is total radial shrinkage, in percent, divided by 100. Neither the initial nor the final moisture content *can be greater than 30 percent. Examples of the* use of this formula to determine dimensional changes are given below:

17

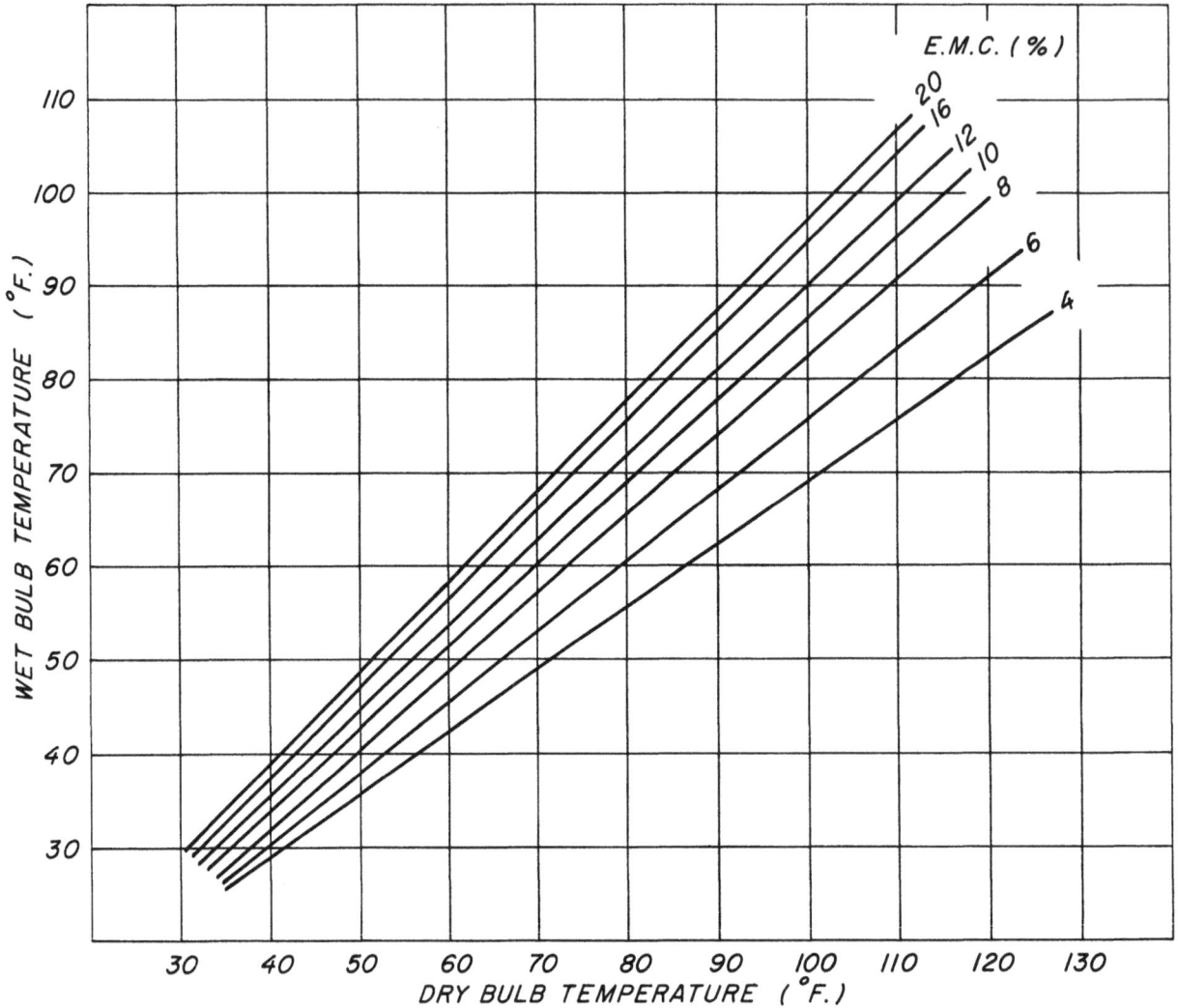

Figure 20.—The equilibrium moisture content of wood as related to the dry- and wet-bulb temperatures of the air.

Example No. 1

Determine the shrinkage in width of a 6-inch, plainsawed, sugar maple board in drying from the green condition to a moisture content of 15 percent. The average green moisture content of sugar maple, sapwood and heartwood combined, is about 69 percent. Since this board is flat grained, use a total tangential shrinkage value of 9.9 percent (table 2). Substituting in the formula:

$$S=\frac{(30-15)6}{\left(\dfrac{30}{0.099}-30\right)+30}=\frac{90}{303}=0.297 \text{ inch}$$

Example No. 2

Determine the shrinkage in width of a vertical-grained, old-growth redwood board, 8½ inches wide, in drying from a moisture content of 24 percent to 5 percent. Table 2 gives the radial shrinkage of redwood as 2.6 percent.

$$S=\frac{(24-5)8.5}{\left(\dfrac{30}{0.026}-30\right)+24}=\frac{161.5}{1147.8}=0.014 \text{ inch}$$

Example No. 3

Determine the swelling in width of a flat-grained, sugar maple flooring strip machined to

18

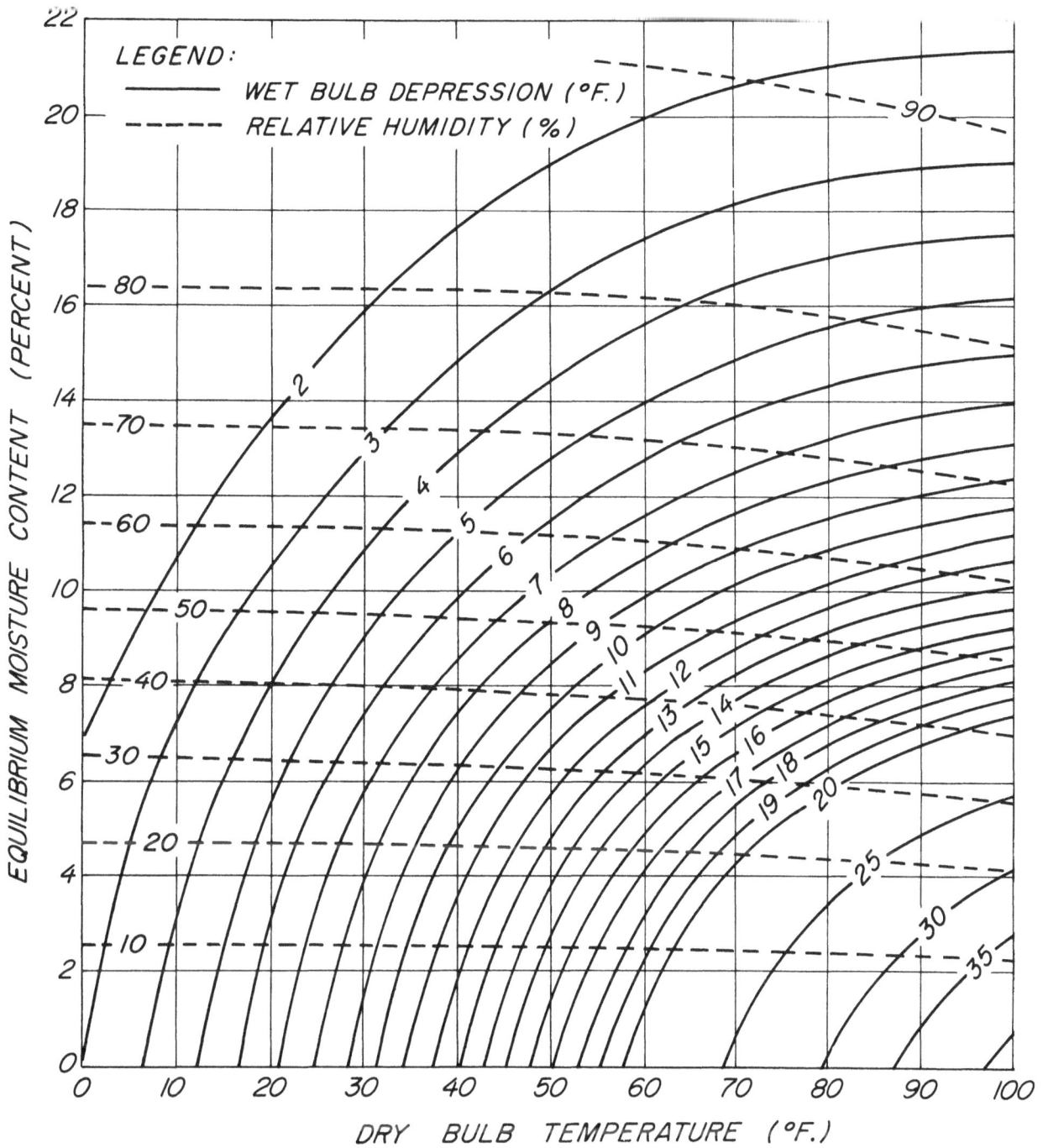

22

LEGEND:

——————— WET BULB DEPRESSION (°F.)

– – – – – RELATIVE HUMIDITY (%)

EQUILIBRIUM MOISTURE CONTENT (PERCENT)

DRY BULB TEMPERATURE (°F.)

M 134 318

Figure 21.—Relative humidity of the air and the equilibrium moisture content of wood as related to the dry-bulb temperature and the wet-bulb depression.

2¼ inches, in changing from 5 to 13 percent moisture content.

$$S = \frac{(5-13)\,2.25}{\left(\frac{30}{0.099}-30\right)+5} = \frac{-18.0}{278} = -0.065 \text{ inch}$$

The negative shrinkage obtained denotes swelling. Using the swelling of 0.065 inch per 2¼ inches of width, a 40-foot-wide floor will swell:

$$\frac{12}{2.25} \times 40 \times 0.065 = 13.8 \text{ inches}$$

19

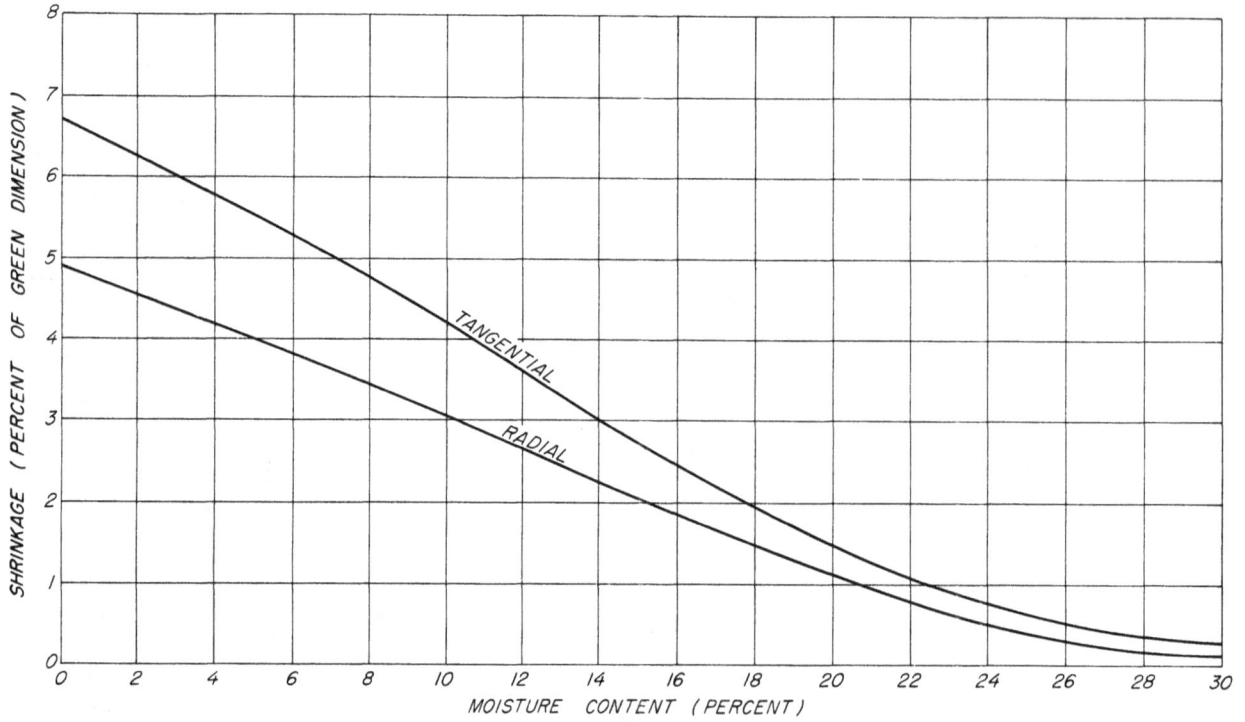

Figure 22.—Typical moisture content-shrinkage curves for the tangential and radial directions of a softwood.

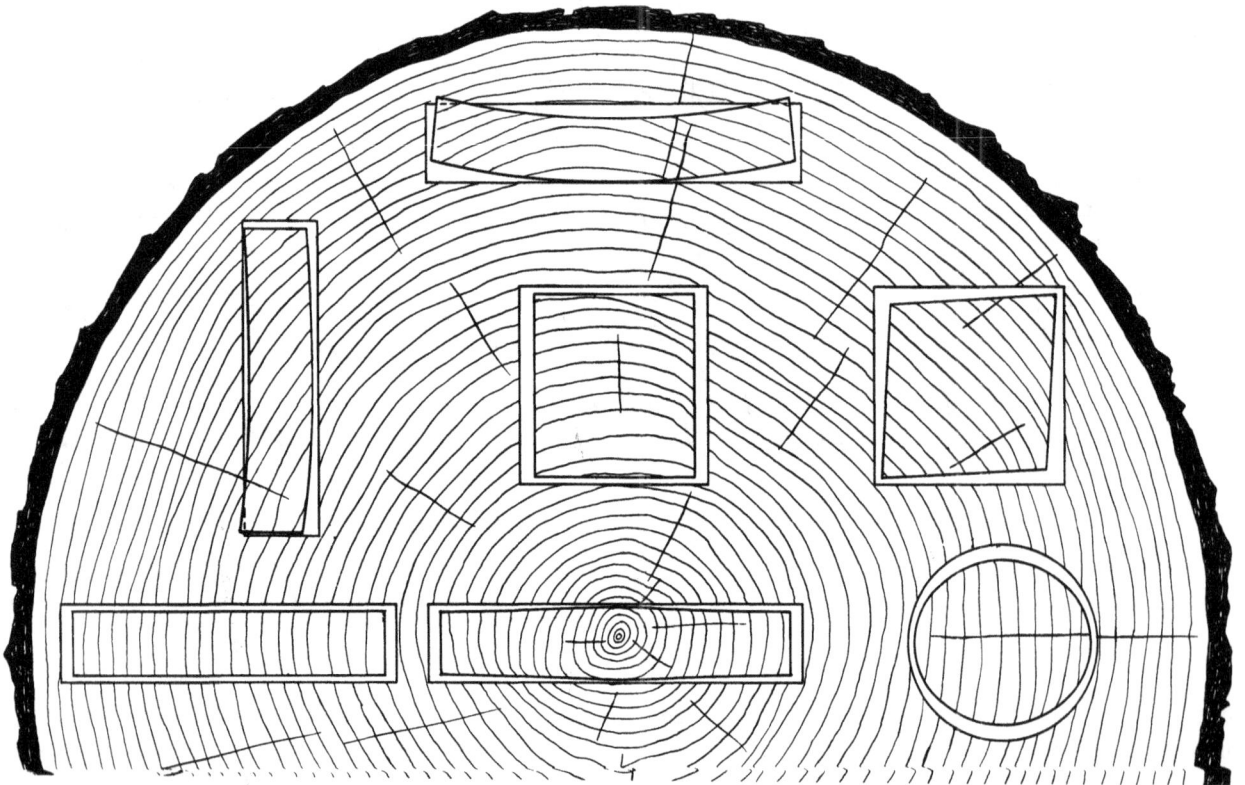

Figure 23.—Characteristic shrinkage and distortion of flats, squares, and rounds as affected by the direction of annual growth rings. The dimensional changes shown are somewhat exaggerated.

20

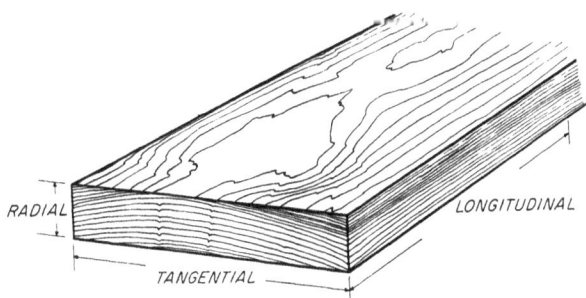

M 136 889
Figure 24.—Shrinkage directions in a plainsawed board.

Weight of Wood

The weight of lumber decreases as its moisture content is reduced in air drying. All wood substance, regardless of the species from which it comes, has an ovendry specific gravity of 1.46. This means that a cubic foot of ovendry wood substance weighs 1.46 times the weight of a cubic foot of water, or $62.5 \times 1.46 = 91.25$ pounds. Usually, the specific gravity of wood is based on the volume of the wood when green and its weight when ovendry. Thus, if the specific gravity of a specimen of green wood is listed as being 0.5, the ovendry weight of the wood substance in a cubic foot of green wood is one-half of the weight of a cubic foot of water, or 31.25 pounds.

The differences in specific gravity between species is due to differences in the size of the cells and the thickness of the cell walls. The average specific gravity, based on ovendry weight and green volume of commercial woods, varies from 0.31 for western redcedar and black cottonwood to 0.64 for the hickories. The moisture in the wood adds to the weight of a given volume. Most species of wood, even when green, float in water because of air entrapped in the cells.

Table 3 gives the specific gravity and the weights of lumber of commercial species per thousand board feet at a moisture content of 20 percent.

Color of Wood

The sapwood of most species is light colored, and can perhaps be described as yellowish white. The heartwood of most species is distinctly darker than the sapwood.

Considerable variation exists in the color of the heartwood of the various species. The colors are generally a combination of yellow, brown, and red. Sometimes a greenish or purplish tint is present. The color of wood exposed to air and light becomes duller and darker. In air-drying yards, it also becomes discolored by wetting and with airborne dust or soot. These naturally caused surface discolorations do not penetrate very far into the wood, except into surface checks, and are removed in planing.

For Additional Information

American Society for Testing and Materials
  1965. Standard methods of test for moisture content of wood. ASTM D 2016–65.
James, W. L.
  1963. Electric moisture meters for wood. USDA, Forest Serv. Res. Note FPL–08. Forest Prod. Lab., Madison, Wis.
Panshin, A. J., DeZeeuw, C., and Brown, H. P.
  1964. Textbook of wood technology. Vol. 1, Ed. 2, McGraw-Hill, N.Y.
Peck, E. C.
  1957. How wood shrinks and swells. Forest Prod. J. 7(7) : 235–244.
Rasmussen, E. F.
  1961. Dry kiln operator's manual. USDA Agr. Handb. 188.
Stamm, A. J.
  1964. Wood and cellulose science. Ronald Press Co., N.Y.
U.S. Forest Products Laboratory, Forest Service
  1955. Wood handbook. USDA Agr. Handb. 72.
Weatherwax, R. C., and Tarkow, H.
  1968. Cell wall density of dry wood. Forest Prod. J. 18(2) : 83–85.

414-042 O - 71 - 2

TABLE 3.—*The average specific gravity and approximate weight of 1,000 board feet of various species of lumber*

Species	Average specific gravity [1]	Average weight of 1,000 board feet [2]	Species	Average specific gravity [1]	Average weight of 1,000 board feet [2]
SOFTWOODS			**HARDWOODS**		
		Lb.			*Lb*
Baldcypress	0.42	3,055	Alder, red	.37	2,669
Cedar:			Ash:		
Alaska	.42	3,055	Black	.45	3,301
Atlantic white	.31	2,248	Green	.53	3,898
Eastern redcedar	.44	3,231	White	.55	4,074
Incense	.35	2,529	Aspen:		
Northern white	.29	2,107	Bigtooth	.38	2,774
Port-Orford	.40	2,915	Quaking	.35	2,529
Western redcedar	.31	2,248	Basswood, American	.32	2,318
Douglas-fir:			Beech, American	.56	4,144
Coast	.45	3,301	Birch:		
Interior north	.45	3,301	Paper	.48	3,512
Interior south	.43	3,161	Sweet	.60	4,460
Interior west	.46	3,371	Yellow	.55	4,074
Fir:			Butternut	.36	2,599
Balsam	.34	2,458	Cherry, black	.47	3,442
California red	.36	2,599	Cottonwood:		
Grand	.35	2,529	Black	.31	2,248
Noble	.37	2,669	Eastern	.37	2,669
Pacific silver	.40	2,915	Elm:		
Subalpine	.31	2,248	American	.46	3,371
White	.37	2,669	Rock	.57	4,214
Hemlock:			Hackberry	.49	3,582
Eastern	.38	2,774	Hickory	.64	4,776
Western	.42	3,055	Magnolia, southern	.46	3,371
Larch, western	.48	3,512	Maple:		
Pine:			Bigleaf	.44	3,231
Eastern white	.34	2,458	Red	.49	3,582
Jack	.40	2,915	Silver	.44	3,231
Lodgepole	.38	2,774	Sugar	.56	4,144
Ponderosa	.38	2,774	Oak:		
Red	.41	2,985	Northern red	.56	4,144
Southern:			Northern white	.60	4,460
Loblolly	.47	3,442	Southern red	.52	3,828
Longleaf	.54	4,004	Southern white (chestnut)	.57	4,214
Shortleaf	.46	3,371	Pecan	.60	4,460
Slash	.56	4,144	Sweetgum	.46	3,371
Sugar	.35	2,529	Sycamore, American	.46	3,371
Western white	.36	2,599	Tanoak	.56	4,144
Redwood:			Tupelo:		
Old-growth	.38	2,774	Black	.46	3,371
Young-growth	.33	2,388	Water	.46	3,371
Spruce:			Walnut, black	.51	3,758
Engelmann	.32	2,318	Willow, black	.34	2,548
Red	.38	2,774	Yellow-poplar	.40	2,915
Sitka	.37	2,669			
White	.37	2,669			

[1] Based on weight when ovendry and green volume.
[2] Weights were calculated for lumber 1⅛ inches thick at 20 percent moisture content. Weights for lumber at moisture content values between 0 and 30 percent can be estimated by using the following formula.

$$\text{Weight at } MC_x = \text{Weight at } MC_{20} + 13\,(MC_x - 20)$$

CHAPTER 3

AIR-DRYING PROCESS

The air drying of lumber involves exposing piles of stickered lumber to the outdoor air. The dry-bulb temperature of the air, its relative humidity as indicated by the wet-bulb depression, and the rate of air circulation are important factors in successful drying. Lumber in the pile dries most rapidly when the temperature is high, the wet-bulb depression large, and air movement brisk through the stickered layers of lumber. The drying rate and the minimum moisture content attainable at any time in any place depend almost entirely on the weather. Air drying, therefore, cannot be a closely controlled drying process.

Utilizing Air Movement

The green lumber dries because air conducts heat to the wood and carries away the evaporated moisture. Thus the air must move within and through the lumber in the pile. As warm, dry air enters the lumber pile, it takes up moisture from the wood and the air temperature drops. As the cool, damp air leaves the pile, fresh air enters and drying continues. The way air moves within a lumber pile depends on the construction of the pile, its location within the yard, and the yard layout and arrangement.

Boards are usually placed edge to edge in the makeup of a unit package and there is little opportunity for the air to drop downward through the package. Consequently, the air must move laterally across the boards. A downward movement of air can be encouraged in piles of lumber by building vertical flues, chimneys, or spacing boards within the pile (figs. 25 and 26). Air enters the pile near the top or from the sides, passes over the broad faces of the boards, cools, and drops downward through the pile openings. A high and open pile foundation permits the moist air to pass out readily from beneath the pile and promotes the general downward movement of air.

The boards in the upper parts of a pile dry more rapidly than those in the lower parts because the top boards are more exposed to wind than the lower ones (fig. 27). The air near the surface of the ground and the bottoms of the piles is generally cooler and consequently has a higher relative humidity than the air near the tops of the piles. The cool, moist air entering the sides of the lumber piles in the lower portions will not induce as rapid drying as the hotter, drier air entering the upper parts of the lumber pile.

In forklift yards, particularly, the orientation of the main alleys or roadways, with respect to the prevailing winds and solar radiation, influences the drying potential of the yard. As air circulation downward through the pile is usually obstructed by edge-to-edge stacking of the boards in each layer, lateral air circulation depends upon air drift stimulated by prevailing winds. If the main alleys are parallel to the prevailing winds and are not obstructed by trees or buildings at the yard edges, the mass of air flowing down the alleys draws air through the lumber piles. If the unit packages are placed parallel to the alleys, the movement of air out of the packages will be parallel to the stickers. Hand-stacked piles of lumber, on the other hand, are usually placed perpendicular to the main alleys and the stickers block air drift parallel to the boards.

Factors That Influence the Drying Rate

The rate at which green lumber will dry after it is placed in the air-drying yard depends upon factors that involve the wood itself, the pile, the yard, and climatic conditions.

Species

Some woods dry rapidly, others slowly. The softwoods and some of the lightweight hardwoods dry rapidly under favorable air-drying conditions. The heavier hardwoods require longer drying periods to reach the desired average moisture content. Specific gravity, then, is a physical property of wood that can guide estimations of drying rates or overall drying time. Southern pine will dry much faster on the yard than will southern oak. On the other hand, the "softer" hardwoods like willow, yellow-poplar, and some of the gums will dry very rapidly. Sugar maple will usually dry faster in northern yards than will northern red oak. Both have about the same specific gravity. The difference here, however, is related to the proportion of sapwood and heartwood.

Thickness

One common rule is that drying time is a function of the square of the thickness. This means

Figure 25.—Unit packages of southern pine lumber stacked with flues to encourage air circulation in the pile.

Figure 26.—Random-width boards in this hand-built pile are spaced on each layer in addition to the chimney to aid circulation within the pile.

that 2-inch stock would require about four times longer to reach, say, 20 percent moisture content, than 1-inch lumber. Actually, drying time varies from a direct proportion of thickness to even longer than the square of the thickness.

Grain Patterns

Quartersawed lumber dries slower than plain-sawed lumber. Wood rays aid the movement of moisture, and in quartersawed lumber few wood rays are exposed on the broad surfaces of the boards.

Sapwood and Heartwood

In softwoods the moisture content of sapwood is usually much higher than that of the heartwood (table 1). However, the sapwood air-dries faster. Usually it will be just as dry as the heartwood, and perhaps drier, when the heartwood reaches the desired moisture content.

In hardwoods, the sapwood moisture content is not often much greater than the heartwood, and is generally lower when air drying is terminated.

Piling Methods

The drying rate of lumber is affected by the way the boards are stacked. For instance, lumber dries faster when air spaces are left between the boards of a unit package than when the boards are placed edge to edge. Flues are occasionally built into wide unit packages to speed up the drying process (fig. 25). Hand-built piles are similarly opened up with flues and chimneys (fig. 26).

24

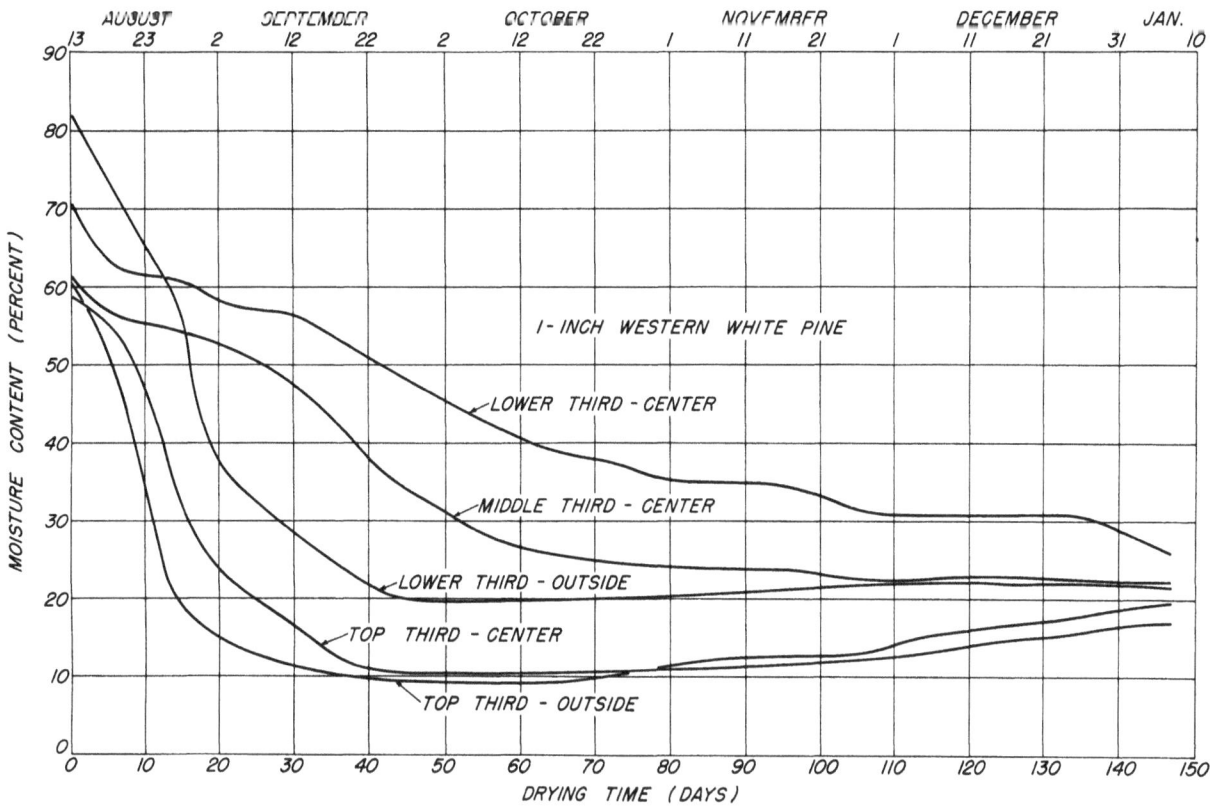

Figure 27.—Moisture content-time curves for the top, middle, and lower portions of a hand-built pile of 1-inch western white pine.

The drying yard can be opened up by spacing the piles farther apart. Increasing the number of alleys is also effective in opening up a yard of hand-built piles of lumber. Placing piles on the outer fringes of the yard will generally increase their drying rate over those placed in the central portion of the yard.

Height and Type of Pile Foundation

While air circulation within the pile is created by natural convection, it is stimulated by winds carrying away the cool moisture-laden air underneath the pile. The foundation under the pile should be reasonably high to allow the prevailing winds to create a brisk air movement under the piles. Pile foundations should not obstruct air flow, nor should weeds or other debris be allowed to block the air passage.

Yard Surface

The drying efficiency of a yard depends to some extent on how well the surface is graded, paved, and drained (fig. 28). If water stands in a yard after a rain, it will decrease the drying rate. The yard should also be kept clean. Vegetation and debris in the form of broken stickers, boards, or pieces of timber from pile foundations (fig. 29) interfere with the movement of air over the ground surface. Vegetation in particular may prevent air from passing out from below the bottom of the piles and interfere with the circulation of air through the lower courses of lumber.

Climatological Conditions

The climate of the area or region in which the air-drying yard is located greatly influences the air-drying rate or yard output of air-dried lumber. Perhaps the most influential factor is temperature, but rainfall is also in the picture. In northern lumber-producing regions, the drying rate is retarded during the winter months by the low temperature. In the southern part of the country, where the winter dry-bulb temperatures are higher, better drying conditions are expected. Unfortunately, these higher temperatures may be offset, for instance in the Southeast, by rains which

25

Figure 28.—Well-graded and surfaced main alleys aid drainage and restrict growth of weeds.

wet the lumber and extend the drying time. In the Southwest, arid conditions can make it difficult to keep degrade losses within bounds.

The average EMC conditions in an area (fig. 30) may not change a great deal from week to week, but in some areas hot, dry winds, or periods of high relative humidity may accelerate or retard drying. Precipitation is particularly important in air drying lumber. Rewetting of the drying lumber

results in a significant slowing down of the drying rate, and the added moisture must be evaporated.

With the conditions of temperature and EMC of the air remaining the same, the drying rate will be relatively rapid when the air movement over the lumber is fast, and slow when the air circulation is stagnant. The orientation of the air-drying yard to the prevailing winds, to effectively stimulate air circulation within the air-drying pile, becomes

Figure 29.—Poor housekeeping results in fallen piles, restricted air circulation, stain and decay, weed growth, and increased fire hazard.

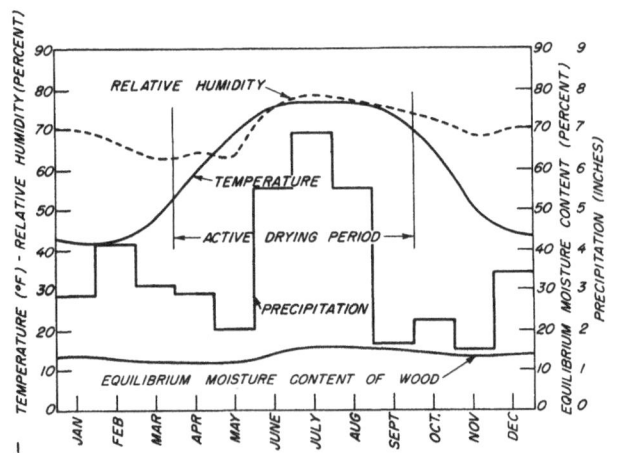

Figure 30.—Average monthly temperatures, relative humidities, precipitation, equilibrium moisture content of wood, and the active drying season in North Carolina, South Carolina, Virginia, and West Virginia.

26

increasingly important as yard output needs to be increased.

Sunshine is also a factor. Solar radiation heats the land areas, exposed areas of the lumber piles, and surrounding buildings. Air moving over these warmed areas and structures is heated by convection, and its drying potential increases. Black bodies absorb more solar energy and become hotter than light-colored materials. This characteristic is used to advantage in air-drying yards by blacktopping the roadways and sometimes the whole area.

In some instances, it may be desirable to arrange the main alleys of an air-drying yard north and south to take advantage of sunshine. The longer sunshine periods on the north-south roadways cause faster snow and ice melt, particularly if these roadways are blacktopped. North-south dirt and gravel roads can also be expected to dry out quickly during the summer months. In regions of high rainfall, faster evaporation of standing water may be a logical reason for locating the main alleys north and south; an example would be the blacktopped yards in the redwood region (fig. 31).

In addition to the general climatic conditions for a region, local climatic conditions influence air-drying results. Yard sites of low elevation, such as near swamps or marshes, or those bordering on bodies of water, are likely to be damper than surrounding areas. Such sites cause slow drying, and encourage the development of mold, stain, and decay. Elevated sites are more likely to be dry. Open sites, in contrast to those surrounded by tall trees, hills, or buildings, are conducive to rapid drying because of the movement of winds through them. One note of warning. A high, dry site encourages rapid drying but may favor surface checking and end splitting of the lumber.

Drying Time and Final Moisture Content

The time to air dry lumber to a predetermined average moisture content depends not only on yard site, yard layout, piling methods, and climatic conditions, but also upon species and thickness. As previously mentioned, lumber of low specific

M 135 049

Figure 31.—A blacktopped air-drying yard with alleys predominately north and south.

27

gravity will dry faster than the heavier woods. The approximate air-drying periods for 1-inch lumber given in table 4 are based on climatic conditions for the region in which the particular species is cut. The data are based on experience with hand-stacked piles that vary in width from 6 to 16 feet. Piles of lumber in unit packages, less than 6 feet wide, would presumably dry in shorter periods.

TABLE 4.—*Approximate time to air dry green 1-inch lumber to 20 percent moisture content*

Species	Time	Species	Time
SOFTWOODS	*Days*	HARDWOODS—con.	*Days*
Baldcypress	100–300	Quaking	50–150
Cedar [1]		Basswood,	
Douglas-fir:		American	40–150
Coast	20–200	Beech, American	70–200
Interior north	20–180	Birch:	
Interior south	10–100	Paper	40–200
Interior west	20–120	Sweet	70–200
Fir [1]		Yellow	70–200
Hemlock:		Butternut	60–200
Eastern	90–200	Cherry, black	70–200
Western	60–200	Cottonwood:	
Larch, western	60–120	Black	60–150
Pine:		Eastern	50–150
Eastern white	60–200	Elm:	
Jack	40–200	American	50–150
Lodgepole	15–150	Rock	80–180
Ponderosa	15–150	Hackberry	30–150
Red	40–200	Hickory	60–200
Southern:		Magnolia,	
Loblolly	30–150	southern	40–150
Longleaf	30–150	Maple:	
Shortleaf	30–150	Bigleaf	60–180
Slash	30–150	Red	30–120
Sugar:		Silver	30–120
Light	15–90	Sugar	50–200
Sinker	45–200	Oak:	
Western white	15–150	Northern red	70–200
Redwood:		Northern white	80–250
Light	60–185	Southern red	100–300
Sinker	200–365	Southern white	
Spruce:		(chestnut)	120–320
Engelmann	20–120	Pecan	60–200
Red	30–120	Sweetgum:	
Sitka	40–150	Heartwood	70–300
White	30–120	Sapwood	60–200
		Sycamore,	
HARDWOODS		American	30–150
Alder, red	20–180	Tanoak	180–365
Ash:		Tupelo:	
Black	60–200	Black	70–200
Green	60–200	Water	70–200
White	60–200	Walnut, black	70–200
Aspen:		Willow, black	30–150
Bigtooth	50–150	Yellow-poplar	40–150

[1] These species are usually kiln dried

The minimum periods given apply to lumber piled during the good drying weather, generally during spring and summer. Lumber piled too late in the period of good drying weather to reach 20 percent moisture content, or lumber that is piled during the fall or winter, usually will not reach a moisture content of 20 percent until the following spring. This accounts for the maximum periods given in the table. For example, 1-inch northern red oak is often air dried to an average moisture content of about 20 percent in southern Wisconsin in about 60 days when yarded in June or July. If piled in the late fall and winter, the stock must stay on sticks for 120 days or more. Local yard and weather conditions and yard layout should be considered as well as the seasonal weather pattern in estimating the periods required for air drying.

Deterioration of Lumber While Air Drying

Losses in lumber value resulting from defects that develop during air drying are reflected in the overall drying cost. Air-drying defects may be caused by shrinkage, fungal infection, insect infestation, or chemical action. Shrinkage causes surface checking, end checking and splitting, honeycomb, and warp. Fungus infection causes blue or sap stain, mold, and decay. Insect infestation results in damage due to pith flecks, pinholes, and grub holes left in the wood. Chemical reactions cause brown stain and sticker marking. If lumber is kept on the yard for an extended period of time, it may appear excessively weathered because of an accumulation of sawdust, ashes, and windborne dirt.

For Additional Information

Clausen, V.
   1954. Air seasoning California redwood. J. Forest. 52:933–935.

Lott, E. J.
   1958. How to pile and season native lumber. Mimeo. F-37, Agr. Ext. Serv., Purdue Univ., Lafayette, Ind.

Peck, E. C.
   1959. Air drying 4/4 red oak in southern Wisconsin. Forest Prod. J. 9(7):236–242.

————
   1953. The air drying of Engelmann spruce. USDA, Forest Serv., Forest Prod. Lab. Report 1944–5. Madison, Wis.

————
   1947. Air drying of lumber. USDA., Forest Serv., Forest Prod. Lab. Report 1657. Madison, Wis.

Smith, H. H.
   1947. Seasoning of aspen. USDA, Forest Serv., Lake States Aspen Report No. 5. North Central Forest Exp. Sta. St. Paul, Minn.

CHAPTER 4

AIR-DRYING YARD

The air-drying yard is generally located near the sawmill producing the lumber or close to the factory using the wood for manufacturing finished products. It is usually laid out for convenient transport. The main roadways or alleys are wide enough for the type of lumber-handling equipment being used (fig. 32). Most yarding operations build up piles of lumber for air drying with prestacked unit packages handled by forklift truck or other mechanized equipment. Overhead cranes or mobile cranes are also used for piling stickered packages of lumber. In some yards, piles are built by hand from either the ground level or from elevated trams or docks.

Yard Layout

Efficient yard layout should provide good drainage of rain and melting snow, free movement of air in and out of the yard, and easy transportation and piling of lumber. A yard laid out for rapid drying potential should be on high, well-drained ground with no obstruction to prevailing winds. However, the need to keep the yard close to the plant limits site selection, and convenient areas are not always favorable to air dry lumber rapidly

and with a minimum of degrade. Yard sites bounded by buildings or with standing water or streams nearby should be avoided, where possible, as lumber drying is retarded.

Most yards are laid out in a rectangular scheme. The alleys or roadways cross each other at right angles, and the areas occupied by the piles are rectangular. Areas may be designated for certain species, grades, or thicknesses of lumber. The alleys serve as routes for transporting the lumber, as pathways for the movement of air through the yard, and as protection against the spread of fire. The alleys in an air-drying yard are classified as main and cross alleys. In large air-drying yards, blocks or areas are often separated by still wider roadways or strips of land to protect the lumber from the spread of fire and to meet insurance requirements. These wide areaways are sometimes called fire alleys.

The sides of the lumber piles in a forklift yard are parallel to the main alleys. Hand-built piles are usually placed perpendicular to the main alley, which might be an elevated tramway (fig. 33). Hand-built pile yards, in addition to the main and cross alleys, also have rear alleys.

Spaces between the sides and ends of the piles are also part of the yard layout. These spaces form additional passageways for air movement.

Alley Orientation

Main alleys in a forklift yard are sometimes oriented parallel to the direction of the prevailing winds. If the lumber on the windward side of a yard is perpendicular to the winds, airflow is greatly retarded. In a yard of hand-built piles, orientation of the main alleys with respect to prevailing winds may not be particularly important if air moves freely underneath high pile foundations.

There are also good arguments for orienting main alleys to take advantage of sunshine. Main alleys dry up more quickly after rain or snow if they run north and south. Where the main alleys run east and west, the piles on the south edge of the alley shade adjacent areas, causing snow to melt more slowly, or the ground to dry more slowly after storms.

M 135 021

Figure 32.—In a forklift yard, the alleys must be wide enough to enable the forklift to travel down the roadway and to maneuver the package onto the pile.

Figure 33.—These piles were built by hand from an elevated tramway.

Alley Size

In a forklift yard, the piles are arranged either in rows or in lines (figs. 34, 35). The main alleys are generally 24 to 30 feet wide. Cross alleys intersect the main alleys at right angles, and provide access to the main alleys. They also afford protection against the spread of fire and may be 60 feet or more in width and spaced every 200 or 300 feet. The alleys, in addition to providing ample room for the forklift truck to maneuver in and out of the rows, must be wide enough to allow clearance for the longest lumber being handled. When carriers are used to transport stickered packages in a forklift yard, the packages are usually set down on one side of the main alley. This means that the alley must be wide enough for the forklift truck to operate in the roadway.

In a yard of hand-built piles, the main alleys are generally 16 to 20 feet wide and 38 to 40 feet apart, or equal to the sum of the lengths of two lumber piles plus the width of the rear alley. The rear alleys should be 6 to 8 feet wide or at least wide enough so that air movement is not blocked by piles of longer length (fig. 36). The rear alleys often have drainage ditches.

Row Spacing

In forklift yards of rows of piles, spaces between the rows should be large enough so the lift truck can operate easily. The rows should be a minimum of 3 feet apart (fig. 34).

In a line-type yard the spacing between lines is usually 2 feet (fig. 37).

30

CROSS ALLEY

50'

24'

3' SPACE BETWEEN ROWS

150'

2'— LATERAL SPACE

CROSS ALLEY

60'

MAIN ALLEY

M 118 854

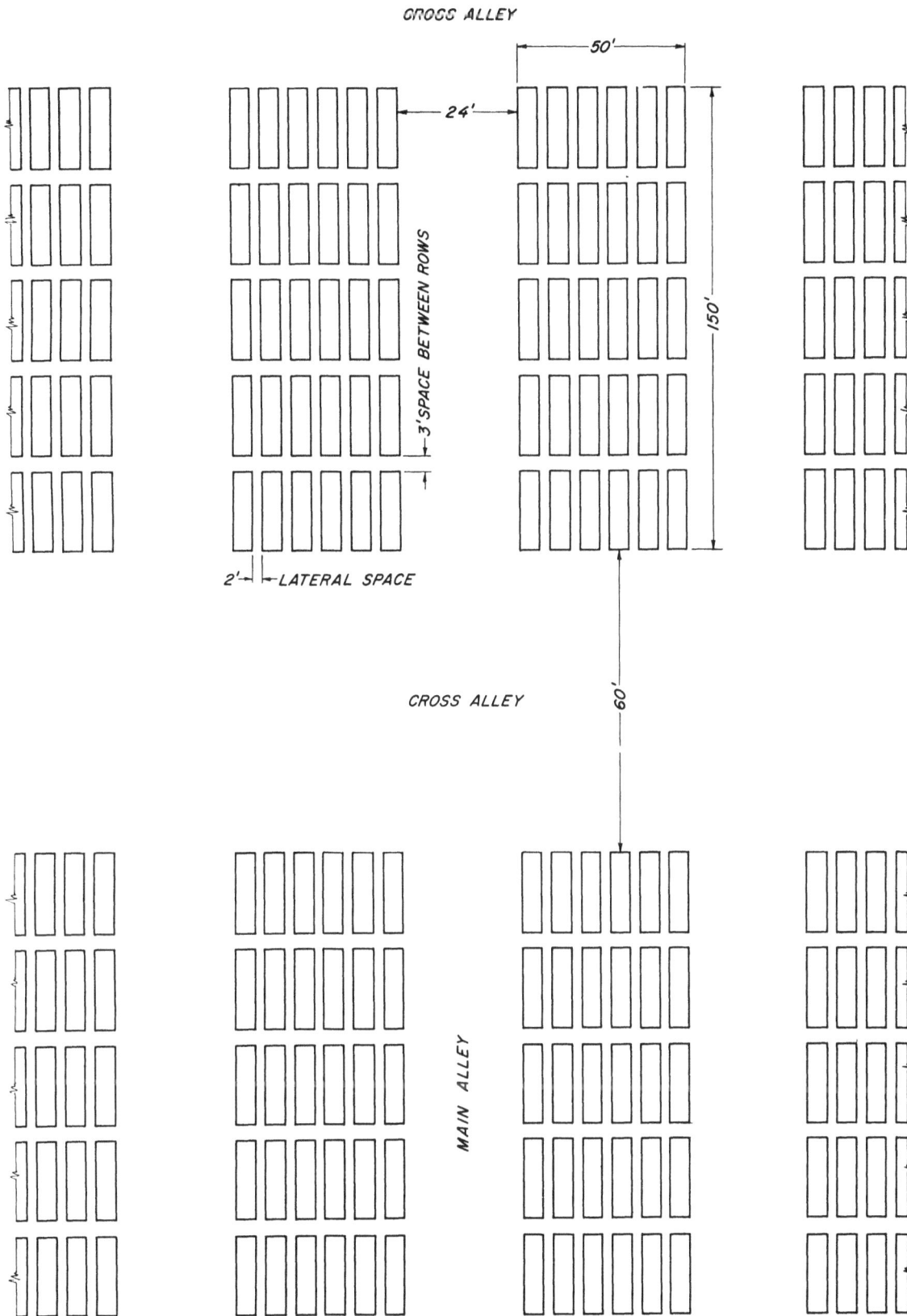

Figure 34.—General arrangement of a row-type forklift yard.

31

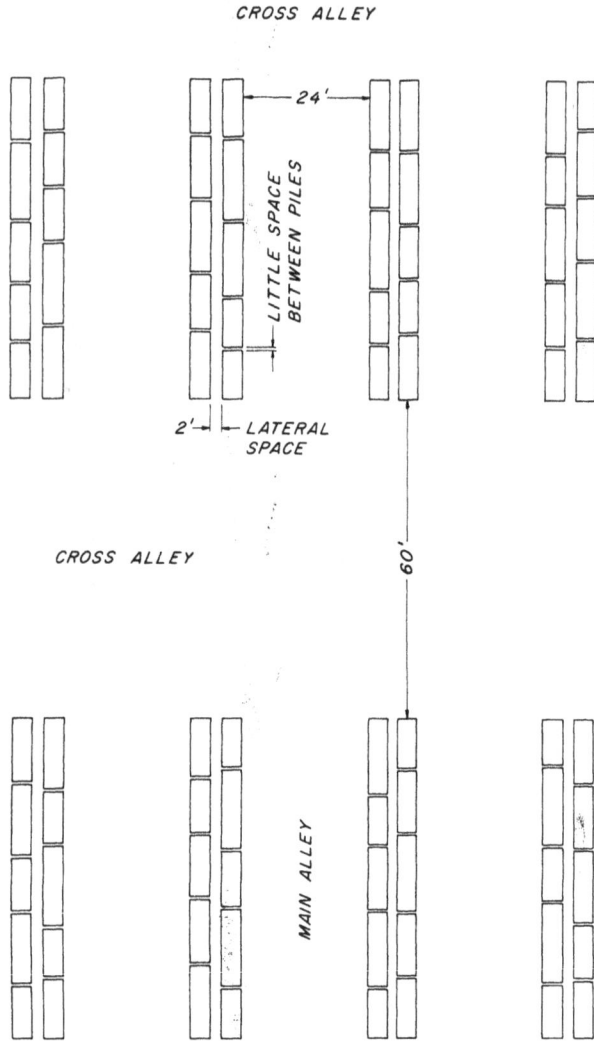

Figure 35.—General arrangement of a line-type forklift yard.

M 1047 F

Figure 36.—Rear alley in a yard of hand-built piles.

M 135 023

Figure 37.—The spacing between the lines in this line type forklift yard is about 2 feet.

Pile Spacing

The spacing of piles varies with differences in climate, yard site, and in the character of the lumber. It also varies with the different specific drying defects to be avoided. Where surface checking is the defect most likely to occur, the width of the spaces should be reduced. Where staining is likely to occur, it is desirable to increase the spacing. In arid regions, during the hot, dry season, piles should be placed closer together than they are during the cool, moist season.

In a row-type forklift yard (fig. 34), the space between the piles within the rows should be about 2 feet but may be as much as 8 feet in the middle of the row when variable spacing is practiced. The

32

spaces between the ends of piles in a line-type fork-lift yard are usually 1 to 2 feet. If the lift truck has a side shifter, the space can be minimized if air movement needs to be reduced. In yards of hand-built piles, the spaces between the sides of the piles should be at least 3 feet.

Pile Widths

For air drying, the width of the lumber piles varies from about 3½ feet to 16 feet. Piles only 3½ feet wide usually consist of unit packages of lumber that are to be placed on kiln trucks after air drying. In yards of hand-built piles, the general pile width for hardwood lumber is 8 or 9 feet, while softwood piles may be as wide as 16 feet.

Pile Heights

Pile heights range from about 4 to 30 feet for both unit-package and hand-built piles. High piles of narrow unit packages are often tied together with long bolsters for lateral support (fig. 38). Thirty-foot-high piles of unit packages may be built by cranes or by forklift trucks. The usual hand-built pile, however, is seldom more than 15 feet high, including the height of the pile foundation. Higher hand-built piles are either constructed from an elevated tramway, or with the help of a machine to elevate the lumber.

Yard Transportation Methods

Unit-Package Yards

The green lumber is transported from the green chain or sorter to a stacker building or to an area where the stickered units are built up. Where the hauling distance from the stacker station to the yard is fairly long, carriers or straddle buggies transport the unit package to the yard site for piling by lift truck. If the hauling distances are short, the slower moving lift truck might both transport the stickered unit package and pile it.

Lumber piles may be built of packages handled by several types of cranes. Tractor cranes operate from the main alleys. Cranes mounted on railway cars operate from tracks laid in the main alleys. Occasionally, lumber piles are built by gantry cranes or monorail cranes. In these yards, the stacked package emerges from the stacker building

M 135 024

Figure 38.—Piles of narrow packages are tied together with bolsters to prevent tipping.

on live rolls or chains to a dock. At the dock the crane or monorail operator picks up the package and transports it to the pile being built up of the particular lumber item. Gantry cranes travel up and down the yard on rails, and build piles of lumber on either side of the rails within reach of their boom (fig. 39). Monorail cranes travel on overhead monorails and build lumber piles beneath the crane.

Hand-Built Pile Yards

A number of methods are used to transport lumber from the green chain or sorter to the air-drying yard in which piles are built by hand. Two-wheeled carts or dollies are still used at relatively small hardwood sawmills. These carts are usually pulled by tractors in the elevated tram or dock-type yard. At larger sawmills, carriers transport the bulk-piled packages of green lumber from the green chain or sorter to the yard. In track-type yards, the green lumber is loaded onto flatcars at the green chain or sorter and then switched by locomotives to the yards.

Influences of Transportation Methods on Yarding

In a forklift yard, the unit packages are already stickered when they are carried out to the yard for piling. Bumping, jarring, and rough handling will displace stickers. The most practical remedial measure is to improve the roadways so that transport can be rapid without jolting the stickered package. Some companies have found it advantageous to blacktop the entire yard surface. Blacktop also absorbs and stores solar radiation and at the same time acts as a moisture vapor barrier against ground water.

In forklight yards where only the main and cross alleys are improved roadways, the initial arrangement of the rows of piles within blocks becomes permanent. The roadway in the rows may be graded and graveled.

In the air-drying yard of hand-built piles, units of bulk-piled lumber are delivered to the piling site by cart, wagon, carrier, or rail. The lumber is bulk piled and can withstand considerable bumping and jarring without falling apart. The

M 86192 F

Figure 39.—In this yard, packages are placed in piles by a gantry crane.

main alleys or roadways for transport in these yards generally are not graded, graveled, or paved to the extent found in lift-truck yards.

Air Drying on Kiln Trucks

Green lumber that is stacked for kiln drying will air dry to some extent in storage before it goes into the kiln. The amount of air drying that occurs depends upon the storage condition. To reduce handling costs, lumber may be stacked on kiln trucks and air dried in a specially arranged track-type air-drying yard (fig. 40). Where air drying and kiln drying are conducted at the same plant, the advantage of air drying lumber on kiln trucks to eliminate takedown and restacking costs may warrant the increased investment in kiln trucks, bunks, and yard tracks. However, this practice was abandoned by many operators when yarding of unit packages with lift trucks was introduced.

Pile Foundations

A pile foundation, or pile bottom, supports the lumber pile and provides clearance between the lumber and the ground. But pile foundations must allow the air which has moved downward through the pile to be readily exhausted. Fixed pile foundations are an integral part of the yard layout, because they determine the location of the piles or rows or lines of piles. Pile foundations represent a considerable capital investment, and they should be well designed and made of materials that will contribute to long life and low maintenance costs.

A pile foundation may consist of the following parts: Mudsills or sleepers, posts or piers, stringers, and crossbeams (fig. 41). Mudsills or sleepers rest upon the surface of the ground or slightly below the surface and support the piers or posts. Wood mudsills should be pressure treated with preservative or be of the heartwood of a decay-

M 135 025

Figure 40.—Hardwood lumber is often stacked on kiln trucks and air dried prior to kiln drying.

35

resistant species. Where the entire surface of the yard is paved with concrete or blacktop, mudsills are unnecessary, and the posts or piers may rest directly on the pavement.

Posts or piers may be made from round or square pieces of wood, concrete, cement building blocks, or masonry. Square wood posts should be about 6 by 6 inches in cross section, and round wood posts should be 6 to 8 inches in diameter. Diagonal bracing may be fastened between the posts to prevent lateral tipping (fig. 42). When posts or piers are set in the ground, they should extend below the frost line and be supported by footings designed to carry the estimated load. This arrangement requires no bracing.

Stringers, running lengthwise of the pile, rest upon the tops of the posts or piers. The stringers may be 6 by 8's placed on edge ,steel I-beams, or railroad rails (fig. 43). Stringers are sometimes eliminated and cross beams rest directly upon the tops of the piers or posts (fig. 44). When this system is used, more piers are needed to support the cross beams because the cross beams are spaced in accordance with the sticker spacing.

Cross beams may consist of wood timbers, steel angle irons, or old rails. If they are made of wood, they should be 4 by 6 inches in dimension and placed on edge. Stringers and cross beams are not so susceptible to attack by decay fungi as mudsills and posts. Stringers, however, may decay where they come in contract with the tops of the posts, or where they come in contact with the cross beams. A thorough brush treatment with a wood preservative at these areas will reduce the decay hazard.

Figure 41.—These permanent pile foundations for sloped, hand-built piles consist of mudsills on the ground, posts between the mudsills and sloped stringers, and crossbeams on which the lower course of lumber rests.

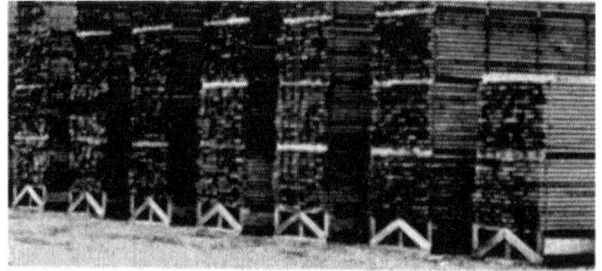

Figure 42.—Braced pile foundations in a row of packages.

Pile foundations should not obstruct or block air movement in any direction. Any arrangement that restricts air movement over the surface of the yard and from beneath the lumber pile, such as laying several long planks or timbers on top of one another to form foundations, is to be avoided. In this instance, air beneath the pile is permitted to move only in the directions parallel to the foundation timbers.

Forklift Yard

Although the forklift yard may place greater reliance upon prevailing winds to create air movement through the packages, the pile foundations should be high enough to allow air drift under the bottom package. Foundations for piles in a forklift yard are usually level and often are movable or temporary. These foundations frequently consist of short timbers, 4 by 4 inches or 6 by 6 inches in dimension, placed directly upon the ground, immediately before building the pile (fig. 45). Sometimes 6- by 9-inch preservative-treated timbers are placed with the broad face on the ground. All timbers except the outer ones are moved aside when the forklift truck travels up and down the space for the row of piles.

Foundations for package piles may consist of frameworks of beams (fig. 46). One set of long beams (or a series of shorter ones) is placed upon the surface of the ground, with enough space between beams for the entrance of the forklift truck. Short blocks and more beams complete the main foundation. With this arrangement, a central support is often used and is put in place just before the pile is built. The framework carrying the crossbeams may be designed to support several piles or a single pile. They may be left in place, like a permanent foundation, or they may be removed.

Another foundation for package piles is the permanent or fixed-beam type. This may consist of a number of fixed beams, the inner pair being at

36

M 135 029

Figure 43.—The railroad rail stringer is supported by concrete piers. Cross beams are placed on the stringers at the sticker tiers.

M 135 028

Figure 44.—Long cross beams for a row of piles in a unit-package yard are supported by concrete piers.

least 9 feet apart to provide an operating space for the forklift truck (fig. 47). Most forklift trucks have a tread of about 8 feet and need a space about 9 feet wide for easy movement in piling and unpiling lumber. The central support located within this space must be removable because the forklift trucks do not have sufficient ground clearance to pass over it. The central support is necessary to prevent sagging of the lumber piles and warp in the dry lumber (fig. 48).

Where the foundations consist of long crossbeams to accommodate a large number of piles in one row, paint marks are often used to indicate

the locations of the piles (fig. 49). This assures uniform lateral pile spacing.

Piles of unit packages are sometimes sloped although this is not a general practice in forklift yards. The slope is accomplished by graduating the heights of the stringers in the foundation (fig. 50) or by sloping the ground. Sloped package piles can also be built by forklift truck if the truck forks can be tilted laterally. If the piles are sloped, the packages themselves should be pitched. If this is not done, then the high end of the pile will be exposed to the weather. Where air drying is followed by kiln drying, pitched packages become

37

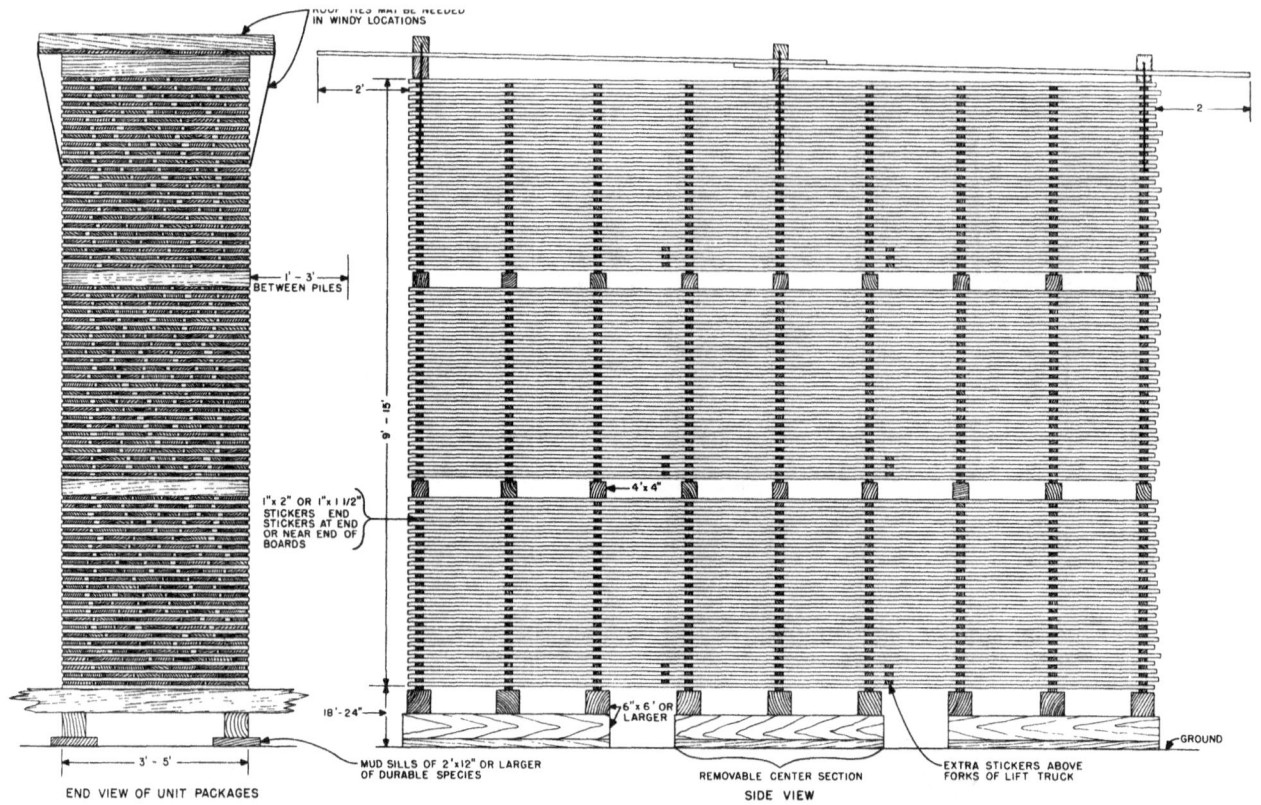

ROOF TIES MAY BE NEEDED IN WINDY LOCATIONS

2'

1'-3' BETWEEN PILES

9'-15'

1"x2" OR 1"x1 1/2" STICKERS END STICKERS AT END OR NEAR END OF BOARDS

18'-24'

3'-5'

END VIEW OF UNIT PACKAGES

4"x4"

MUD SILLS OF 2"x12" OR LARGER OF DURABLE SPECIES

6"x6' OR LARGER

REMOVABLE CENTER SECTION

SIDE VIEW

2

GROUND

EXTRA STICKERS ABOVE FORKS OF LIFT TRUCK

M 134 841

Figures 45.—The diagram illustrates the features of a level pile of unit packages and the design of all-wood movable foundations.

unworkable, because kiln truckloads cannot be built readily from pitched packages.

Hand-Built Pile Yard

Wood foundations for flat piles built by hand often slope from front to rear of the pile (fig. 51). The slope is usually about 1 inch per foot of length. If a pile is sloped, the foundations should be sufficiently high to raise the lumber 18 inches above the ground surface at the low or rear end. For a 16-foot pile, this causes the front end to be 34 inches above the ground. The pile foundation may accommodate a single pile only, or a row of piles. The latter type is composed of a series of long, continuous cross beams (fig. 52).

Lumber Pile Protection

One of the disadvantages of yard drying is exposing lumber to the weather. Exposure to direct sunshine and rain or melted snow causes alternate drying and absorption of moisture. Roofs and covers of various types can protect lumber while it dries and avoid degrade and loss in value.

Pile Roofs or Covers

An effective pile roof is an essential feature of good airdrying practice. A roof protects the upper courses of lumber and, to a lesser extent, the lower parts of the pile from direct sunshine and precipitation. Without a roof the lumber in the upper courses, and particularly the top course, may warp, check, and split. Rain or snow penetrating the pile may retard drying, contribute to the development of fungus stains and chemical sticker stain, and cause surface checks to increase in size. A pile of low-grade lumber, however, may not necessarily justify a roof because of its low value and the improbability that it will drop in grade because of exposure. Some plants that do not use pile roofs reduce the amount of damage by placing a package of low-grade lumber at the top of each pile.

To afford maximum protection, a roof should project beyond the ends and sides of the pile. For a level pile of unit packages, the roof should project about 2 feet at both ends (fig. 53). Roofs on packages do not usually project on the sides, because the roof would be in the way of the forklift.

38

Figure 46.—Long beams are separated by short timbers to provide an open foundation for unit package piles. The center support for the pile is readily movable.

A sloped roof will permit the water to drain off (fig. 54).

A roof for hand-stacked, sloped piles should project about 1 foot at the front, 2½ feet at the rear, and even a 6-inch side projection will be beneficial (fig. 55). The extension over the rear end of the pile allows the drip to fall free. The slope of the roof on such a pile should follow the slope of the pile, if the roof is reasonably tight. If the roof is not tight, the slope should be increased to about 1 inch in 9 rather than the normal slope of 1 inch in 12.

Various methods are used in constructing and placing pile roofs. Loose-board roofs are commonly made from low-grade lumber. Roof boards may be used repeatedly or sold after several uses. The boards may be placed in single layer, single length; double layer, single length; and double layer, double length.

Loose-board roofs on package piles are generally not sloped enough to provide good drainage.

Piles of lumber may be roofed with waterproof paper, building paper, or roll-roofing laid directly on the top course of lumber and weighted with low-grade lumber (fig. 56). These materials may also be combined with boards to form a panel roof (fig. 57). A pile roof may consist of a single panel or a pair. A single panel is usually designed to slope from one end of the pile to the other. Where two panels are used, they overlap at the center and slope toward both ends. Paper or roofing felt provide watertightness while the boards support the paper or roofing in a flat sheet and permit the panel roof to be anchored to the pile. The boards

39

Figure 47.—The spacing between the inner pair of these permanent foundations allows the forklift to move easily down the row. The center support is removable.

M 98620 F

M 135 032

Figure 48.—Unit-package piles on built-up beam foundations with center supports to prevent sagging of the lumber. A short tier of stickers between the main tiers stiffens the package at the points of lift.

in this type of roof should be laid edge to edge in a single layer. A wide variety of building papers and roofing is available and the choice of a suitable material should be based on the cost and the expected life of the material. Panels may also consist of boards nailed to a light wood framework of 2 by 4's or 2 by 6's with battens over the joints.

Panel roofs for package piles are usually fastened to the top package while it is still on the ground. The panels may be handled manually or by a forklift truck.

Roofs should be secured against wind. Tie pieces of 2- by 4- or 4- by 4-inch lumber can be laid across the front, middle, and rear of the roof, and fastened by wires or springs to the pile about 10 courses below the top. A panel-type roof can be secured by fastening a wire or rope from the panel edge to a stick inserted in the side of the pile (fig. 58).

The narrowness of package piles encourages the use of numerous materials in the design of pile roofs. Exterior-grade plywood, or hardboard, may be used. Sheets of galvanized corrugated iron and corrugated aluminum are also satisfactory. Where loose sheets are used, they are supported by cross pieces placed on the top of the lumber pile, and held down by clamps (fig. 59).

Sun Shields and Baffles

Pile roofs that only project 1 to 2½ feet at the ends of a pile do not protect or shade the ends of all the boards from direct sunshine. Piles of high grade or high value lumber may need the protection of sun shields to keep the lumber from drying too fast. Prefabricated lightweight panels can be fastened to the end of the unit package before it is placed on the pile (fig. 60). On hand-built piles, boards arranged like louvers are sometimes attached to the pile end to shield the board ends from direct sunshine (fig. 61). Shielding the board ends reduces the tendency to end check and split under the heat of the sun. In some instances, the sides of piles of thick lumber exposed to direct sunshine are boarded up to minimize edge checking.

Refractory woods exposed to hot, dry winds may surface check badly. By baffling the air movement through the stickered lumber, the surface EMC conditions are modified enough to prevent excessive surface checking. Canvas has been used for pile baffles but plastic tarpaulins are effective and much easier to handle.

End Coatings

Moisture-resistant coatings are sometimes applied to the end-grain surfaces of green lumber to

40

M 98792 F

Figure 49.—Permanent foundations of long crossbeams support the unit packages at the outer sticker tiers. Paint marks are guides for locating piles in the row. Removable center supports are stored between the beams.

retard end drying and minimize the formation of end checks and end splits (fig. 62). The wood beneath the coating is maintained at a higher moisture content than if the coating were not used. To be effective, the coatings must be applied to the freshly trimmed green lumber before any checking has started.

Many kinds of end coatings are commercially available, but are basically of two types—those that are applied cold, and those that are applied hot. Cold coatings are most widely used for lumber products. They are applied by swab, brush, or spray.

It is important to obtain a thick coating; if the material is thinned for easier application, additional coatings may be necessary. The cost of an end coating and its application must be justified by the value of the lumber saved. It is more profitable to apply end coatings to high-grade lumber than to low-grade lumber, and to 2-inch and thicker boards. Nominal 1-inch lumber, for example, is seldom end coated.

Shed Drying

Lumber, particularly the higher and more valuable grades, is sometimes air dried in open sheds (fig. 63). The air-drying shed is essentially a permanent roof, so the lumber is not rewetted by rain. In areas or regions where rain wetting unduly extends the drying time in the conventional air-drying yard, shed drying reduces the time required to attain the desired moisture content.

The sheds are generally open so that the sides and ends do not obstruct air movement. Under certain conditions, however, they may be latticed or louvered. For instance, thick, refractory woods are often air dried in sheds with closed sides as a means of reducing surface checking.

M 135 083

Figure 50.—The height of the concrete piers is graduated to produce a sloped stringer. The crossbeams are placed at the sticker tiers. The unit packages are pitched.

41

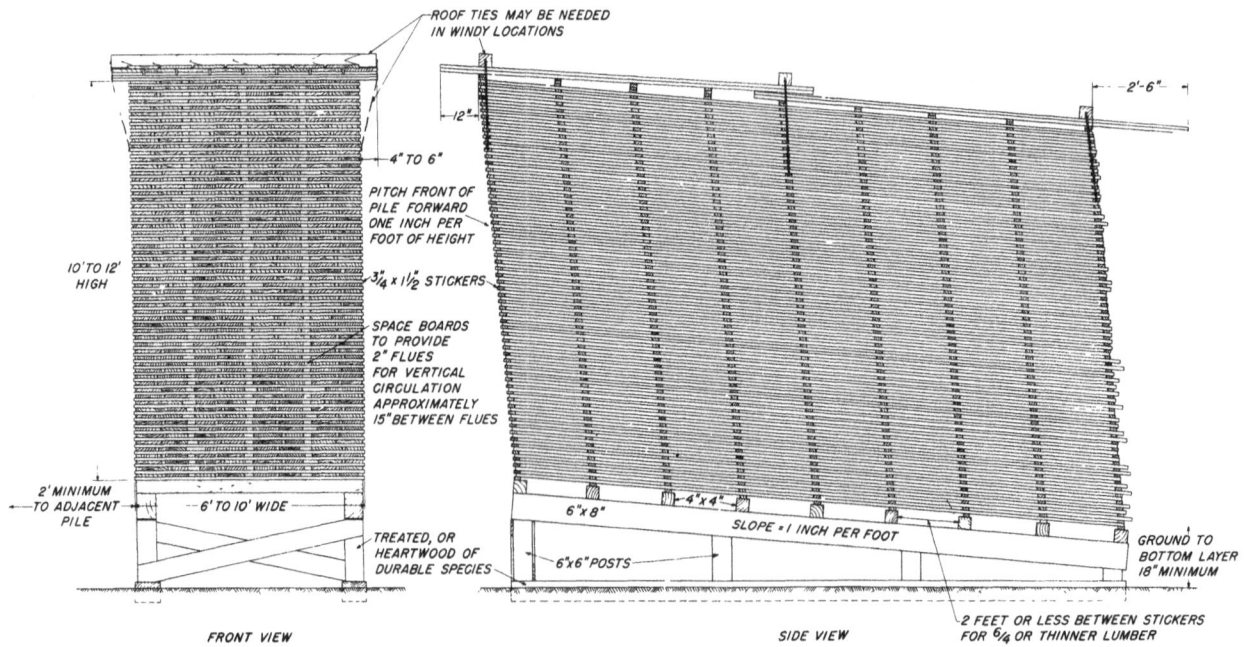

ROOF TIES MAY BE NEEDED IN WINDY LOCATIONS

12"

2'-6"

4" TO 6"

PITCH FRONT OF PILE FORWARD ONE INCH PER FOOT OF HEIGHT

10' TO 12' HIGH

$3\frac{3}{4}$" x $1\frac{1}{2}$" STICKERS

SPACE BOARDS TO PROVIDE 2" FLUES FOR VERTICAL CIRCULATION APPROXIMATELY 15" BETWEEN FLUES

2' MINIMUM TO ADJACENT PILE

6' TO 10' WIDE

TREATED, OR HEARTWOOD OF DURABLE SPECIES

6"x 8"

4" x 4"

SLOPE = 1 INCH PER FOOT

GROUND TO BOTTOM LAYER 18" MINIMUM

6"x6" POSTS

2 FEET OR LESS BETWEEN STICKERS FOR $^6/_4$ OR THINNER LUMBER

FRONT VIEW

SIDE VIEW

ZM 72541 F

Figure 51.—The diagram illustrates the features of a sloped, hand-built pile and the design of an all-wood foundation.

M 44084 F

Figure 52.—Permanent foundations for hand-built piles of softwood lumber. The cross beams supporting the piles at the sticker tiers are supported by concrete piers.

42

Figure 53.—These prefabricated panel roofs extend beyond the ends of the unit packages.

Figure 54.—In this forklift yard, the prefabricated board and batten roof is sloped to provide drainage.

Shed-dried lumber is usually brighter than lumber air dried on a yard because weathering or discoloration due to rewetting is prevented. As the shed roofs usually extend beyond the piles and protect them from sunshine, end checking and splitting are also greatly reduced. Another advantage of shed drying is that air-dried stock can be stored there without deterioration if space is available (fig. 64).

Lumber-drying sheds are usually loaded by lift trucks. Pile-foundation requirements are the same as for the forklift yard. The shed floor may or may not be paved. If not, grading may be required, de-

Figure 55.—A double-layer and double-length, loose-board roof placed on a sloped and pitched, hand-built pile in a dock-type yard. The roof projects about 1 foot at the front and about 2½ feet at the rear.

43

M 136 920

Figure 56.—Reinforced waterproof paper is used as a pile roofing material and weighted with low-grade boards.

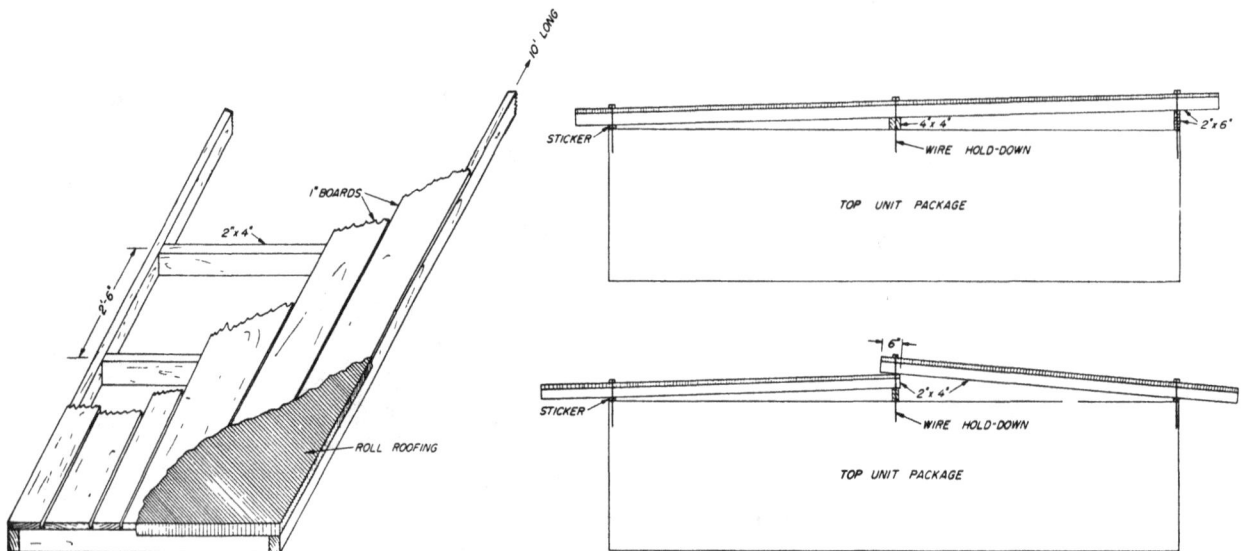

ZM 92586 F

Figure 57.—Roll roofing is used to form a watertight panel roof of boards and framing.

M 121 574

Figure 58.—The prefabricated panel roof on these packages is tied to the top packages with wire or rope.

M 135 039

Figure 59.—Loose sheets of corrugated metal are tied to the package with "C" clamp. A sticker is laid across the top of the package on which the clamp is placed.

M 135 041

Figure 60.—Sun shields are prefabricated by nailing or stapling thin lumber to cleats. The shields are tied to the ends of the package with wire.

M 98839 F

Figure 61.—Portable sun shields on the ends of a hand-built pile of thick lumber.

pending upon soil conditions, and perhaps gravel or crushed-stone application can be justified.

The sheds may be fairly wide to make up long rows of piles within the bays. Entry to the rows would usually be from both ends of the rows or both sides of the shed. On the other hand, the sheds may be long and narrow with two lines similar to a line-type forklift yard (fig. 63).

Drying sheds are usually pole-type structures although such other structural materials as steel posts and metal trusses and roofing may be used. Where increased air circulation is desired, orientation of sheds parallel to the prevailing winds seems logical to stimulate air movement through the stickered unit packages.

Yard Operation and Maintenance

Supervision of an air-drying yard involves considerable responsibility and skill. The yard superintendent is often required to lay out new air-drying yards and expand or extend old ones. Conversion from a yard of hand-built piles to unit-package handling and piling may also be involved.

46

M 135 042

Figure 62.—This lumber is being end-coated after the piles are built up. Sprayable wax emulsions are used. End coating is used instead of sun shields.

Site, orientation of main alleys, sizes of main and cross alleys, length of lines or rows, row spacing and pile spacing within rows and lines, and size of blocks all require consideration for the inventory of lumber anticipated.

The roadways in an air-drying yard are often busy highways for several kinds of transport trucks. Grading, graveling, or crushed-stone fill or paving are investments to improve the mobility of wheeled equipment. Frequent inspections of road conditions are a responsibility of the yard superintendent. In northern areas, snow-removal equipment is needed to keep the main alleys open for yard traffic.

The design, installation, and repair of pile foundations require considerable attention. Sagging foundations cause warped lumber and pile tipping. Sometimes the yard superintendent is also re-

47

Figure 63.—Unit packages are piled on permanent foundations in this drying shed.

sponsible for sorting and unit-package stacking at the sawmill. Sorted-length stacking or box piling involves decisions about sticker spacing and sticker alinement. Hand-built piles may require use of a frame of sticker guides when less-skilled stackers are piling the lumber. Automatic or semiautomatic stacking equipment requires routine inspection and maintenance.

Investment in stickers justifies careful attention to their fabrication and storage. Dry storage is recommended to minimize infecting green lumber with stain and decay fungi. Excessive sticker breakage requires a determination of causes and corrective measures. Sticker replacement costs may justify salvage operations.

The design, fabrication, handling, and storage of pile roofs or covers and sun shields are a yarding operational task. Reduced degrade and footage losses can make pile protection pay off.

Weed control and preventing the accumulation of debris that creates fire hazards are other func-

tions of yard operational supervision (fig. 65). Fire-prevention policies and regulations must be formulated and routine inspections made for compliance.

As air drying is a moisture content reduction process, check tests with portable electric moisture meters enable the yard supervisor to determine when lumber in certain blocks or areas in the yard is ready for removal. As yarding costs not only involve time on the yard but grade and footage losses as well, the yard supervisor should conduct degrade studies to determine how and where yarding practices should be changed to significantly reduce these losses.

Records of the lumber placed in an air-drying yard must be kept. An inventory system best suited for the operation is essential. The location of lumber in the yard by species, thickness, grade, footage, and placement time must be readily determinable. Unit packages or hand-built piles should be ticketed to show date of piling, as well as the

48

Figure 64.—The open shed can be used for storing air-dried lumber when the space is available.

species, thickness, length, grade, and footage of the lumber; duplicates should be systematically filed in the yard supervisor's office. Package or pile tickets are sometimes placed on a large panel on which the air-drying yard is outlined, showing where the lumber is located. Plants using electronic computerized cost-accounting systems or automatic data-processing equipment probably would use the equipment for inventory control purposes.

M 135 046

Figure 65.—Rank weed growth in a terraced air-drying yard restricts air circulation and increases fire hazards.

For Additional Information

Baker, G., and McMillen, J. M.
1955. Seasoning beech lumber. Northeastern Technical Committee on Utilization of Beech in cooperation with the USDA, Forest Serv., Beech Utilization Series 11, Northeast. Forest Exp. Sta., Upper Darby, Pa.

Clark, W. P., and Headlee, T. M.
1958. Roof your lumber and increase your profits. Forest Prod. J. 8(12) : 19A–21A.

Cuppett, D. G.
1966. Air drying practices in the central Appalachians. USDA, Forest Serv., Northeast. Forest Exp. Sta. Res. Pap. NE 56, Upper Darby, Pa.

Keer, G. A.
1955. Air seasoning. Wood 20 : 363–365.

Mathewson, J. S.
1930. The air seasoning of wood. U.S. Dep. Agr. Tech. Bull. 174.

Page, R. H.
1957. Protection of lumber with stack covers while air drying. USDA, Forest Serv., Southeast. Forest Exp. Sta. Release 12, Asheville, N.C.

Page, R. H. and Carter, R. M.
1958. Variations in moisture content of air seasoned southern pine lumber in Georgia. Forest Prod. J. 8(6) : 15A–18A.

Peck, E. C.
1952. New look in air-drying lumber yards. South. Bldg. Supplies. Apr.

————
1952. Machines take over the lumber yard. Nat. Assoc. of Commission Lumber Salesmen Yearbook.

————, Kotok, E. S., and Mueller, L. A.
1956. Air drying of ponderosa pine in Arizona. Forest Prod. J. 6(2) : 88–96.

Rietz, R. C.
1965. The air drying of southern hardwoods. South. Lumberman 210(2617) : 19–20 (May 1).

Skolmen, R. G.
1964. Air drying of Robusta eucalyptus. USDA, Forest Serv., Pacific Southwest Forest and Range Exp. Sta. Res. Note PSW 49, Berkeley, Calif.

Whitmore, Roy
1958. Air-drying lumber to increase mill profits. USDA, Forest Serv., Central States Forest Exp. Sta. Tech. Pap. 158, St. Paul, Minn.

CHAPTER 5

PILING METHODS FOR AIR DRYING

To prepare lumber for air drying, it is usually laid up into courses or layers with separating stickers (fig. 66). The objective is to expose the board surfaces throughout the pile to air circulation. In a forklift yard, the stickered unit packages are made up at a stacker location. In a hand-built yard, a skilled lumber stacker makes up the pile in the yard. The unit packages or hand-built piles are often made up of lumber previously sorted by species, size, and grade.

Sorting of Lumber for Air Drying

Lumber coming from the sawmill is generally sorted into classifications based on similar drying characteristics before it is sent to the unit-package stacker or to the yard for stacking on hand-built piles. Sorting facilitates the stacking operations.

Sorting for Species

Although a number of softwoods and hardwoods have quite similar drying characteristics,

M 134 963

Figure 66.—The lumber in these stacked packages is separated by stickers to allow air circulation over the board surfaces.

the species sort is usually made for merchandising purposes. Those woods which dry rapidly without serious degrade can be yarded in areas where drying conditions are favorable. Species that dry slowly and are likely to surface check can be air dried in yard areas where drying conditions are less severe.

Sorting for Thickness

The time required to air dry lumber to a predetermined moisture content is greatly influenced by board thickness. For example, 2-inch lumber may require three to four times longer than 1-inch lumber of the same species. It is the usual practice, therefore, to segregate the rough-sawed lumber by thickness classes for air drying. Miscut lumber is difficult to stack and thinner parts of the board cannot be held flat. One thick board in a course may allow adjacent well-manufactured boards to warp because of lack of restraint. Presurfacing miscut lumber to a uniform thickness facilitates stacking and reduces warp, sticker breakage, and sticker deformation.

Sorting for Width

Softwoods are generally sawed in the mill to produce certain final dry dressed sizes. These width classes are kept together for drying to reduce sorting and rehandling costs after drying (fig. 67). Hardwoods are most often sawed to random widths in the sawmill, and width segregation for air drying is seldom practiced.

Sorting for Length

Good unit-package buildup by hand or by mechanical stackers is best accomplished by sorted-length stacking (fig. 68). The main objective is to gain as much restraint to distortion as possible through good stickering practices. Hand-built piles, too, are erected more easily if lumber of the same length is used. The advantages of sorting for length apply to both softwoods and hardwoods. Redwood is one of the few species that can tolerate the unsupported ends of mixed-length stacking without serious losses due to warp.

Sorting for Grain

A plainsawed board dries faster than a quartersawed board, and mills specializing in producing quartersawed lumber may find it advantageous to sort for this characteristic. Plainsawed lumber is more susceptible to surface checking than quartersawed lumber and is often yarded in areas where the drying potential is less severe. When relatively few quartersawed boards are produced along with the major production of plainsawed boards, they are seldom segregated.

Sorting for Grade

The separation of green lumber by grades is generally a matter of keeping like-value stock together. Higher grades of lumber may be given bet-

M 134 964

Figure 67.—The lumber in these unit packages was sorted for width, thickness, and length.

Figure 68.—Piles of packages that were sorted for thickness and length.

ter protection in the piles or may be shed dried. The stickering practices may involve more tiers of stickers to reduce degrade caused by warp. The moisture content quality standards of the air-dried upper grades of lumber are often more exacting and require longer air-drying periods than the lower grades.

Sorting Equipment

A number of sorting methods have been developed to segregate lumber into various classifications for air drying.

Conventional Green Chain

Sawmill lumber items are separated into the predetermined sorts by pulling the pieces from a conveyor chain which carries the lumber out of the mill (figs. 69, 70). The lumber is bulk-piled in units for transportation to the stacker by carrier, forklift truck, crane, or transfer chains. Sometimes the lumber is stacked and stickered in bins alongside the green chain, particularly if the unit packages are small, as for forklift truck operation (fig. 71). Most often, however, the units on the

green chain are solid-piled handling units that are transported elsewhere for stacking. The conventional green chain at a sawmill, large or small, requires a high percentage of the manpower, particularly if the sawmill itself has been modernized.

Edge Sorter

This equipment usually consists of a long continuous row of live rolls with a number of slots into which the operator feeds the boards for the various length, grade, and width combinations (figs. 72, 73). An improved sorter, in which the number of slots is reduced, separates the lumber by 2-foot intervals by electrical or mechanical devices (fig. 74). A two-slot edge sorter, for example, may have 4/4 lumber fed into one slot and 8/4 lumber put down on the other side with the boards electrically or mechanically separated into all length and width combinations. Lumber is manually stacked at the sorter or is transported to a mechanical stacker. In manual operations, two men move as a team from bay to bay to remove the lumber from the bin and place it on the stack, which may or may not have sticker guides. Mechanization can also be provided; when a full package of lumber has collected, an operator can

53

empty the bay by pushing a button to activate the chains in the bottom of the bay. The lumber is automatically fed to a transfer belt that carries the stock to the stacker.

Pocket Drop Sorter

In the pocket drop sorter, lumber is transferred crosswise and carried on an overhead lug chain. The boards are dropped into dump buggies according to length, width, thickness, and grade. The dump buggies move on rails (fig. 75); full buggies are transferred to a buggy dumper that feeds a mechanical stacker.

The boards may also be dropped into bins rather than buggies. When the bin is full, an operator can feed the boards onto a transfer chain located under the bins. This transfer chain feeds the mechanical stacker.

Tray-Type Sorter

This device automatically sorts lumber into trays and carries the lumber to the end of the tray (fig. 76). The trays can accommodate enough lumber to make up a full unit package. Lumber is conveyed down these trays by chains, or the trays can be set on a decline and equipped with skate rolls so lumber will travel to the tray end by gravity. The boards are fed into the trays from an inclining chain with lugs; it is provided with either a mechanical or electrical signaling device to eject the board into the proper tray. This type of sorter provides extreme versatility as the various trays can be used for different lengths, grades, thicknesses, or widths as production requires. The tray-sorter system has a limit of about 15 openings and the distance required to hold a full package is rather long. Normally, it is necessary to have sufficient room for at least a package and a half to insure that the tray will not become full before the stacker can empty the tray. The stacker operator unloads the trays by transfer chains, transporting the lumber to the stacker layup table.

One factor that limits sorting is the time required to accumulate enough of any particular item. If the sorting is carried too far, the lumber of each kind accumulates so slowly that it takes too long to complete a pile. Lumber in the lower

M 134 967

Figure 69.—Hardwood lumber is graded on this conventional green chain and the sorts are pulled and placed on dollies. The dollies are rolled out for forklift pickup and transported to the yard or to the unit-package stacker.

54

M 134 968

Figure 70.—Lumber is pulled off this cable-type green chain onto piles that are transported to the stacker by forklift truck.

parts may become quite dry before the pile is completed. This problem is generally more critical when lumber is stacked in hand-built piles, because the piles usually are wider and contain more lumber than unit-package piles. Consequently, it is possible to sort to a finer degree when lumber is made into packages and piled by forklift than in hand-built piles. In addition, piles of packages do not necessarily have to be composed of a single item. For example, the three packages of a pile may all contain the same species and thickness, but different grades.

Stacking Lumber in Unit Packages

Unit packages are often stacked by hand, using a crew of one or two men (fig. 77). Packages are made up beside the sorter or at a special stacking location to which the sorted lumber is delivered.

Because the stacking of unit packages can be done at one place, in contrast to stacking lumber onto hand-built piles in the yard, a stacker building is often constructed to shelter the workers. The stackers stand on the ground or on a platform that is approximately at the same level as the bottom of the package of lumber. As the stickered packages are seldom over 4 feet high, the whole package can be constructed without much lifting of the boards. The packages are usually built up on a stacking rack or jig equipped with sticker guides. The sorted lumber is delivered to the stackers in bulk units, which are placed alongside the stacker frame. The completed stickered package is lifted out of the sticker guide framework by forklift truck, transported to the yard, and placed on a designated pile.

Where the volume of lumber being stacked for air drying is sufficient to justify their operation, automatic or semiautomatic stackers are sometimes installed. An automatic stacker is identified by the magazines that hold the stickers. The stickers are mechanically fed out of the bottom of the magazine onto the lumber layers. A sticker magazine is located at each sticker tier, and the magazines are kept filled by a stickerman. The semiautomatic stacker lays up the courses of lumber onto the package but the stickers are laid manually into sticker guides on both sides of the stacker frame (fig. 78).

The lumber is transferred from the sorter as bulk units that are set on a conveyor, either roller or chain, that moves the unit to a breakdown hoist. The boards move from the breakdown hoist to an "unscrambler" where they are separated into a single layer and moved by conveyor to the stacker layup table. A grading and tally station might be located between the breakdown hoist and the stacker layup table.

When a course of lumber has been laid on the package by the stacker arms and the stickers laid automatically or manually, the package is lowered mechanically so that it is in a position to receive the next course. When the package is completely stacked, it is removed from the stacker and conveyed on rollers or chains to a station where a lift truck or a carrier can pick it up for transport to the air-drying yard.

Stacking Method

The stickered unit packages can be built up from sorted-length lumber, box piled, or stacked with both ends ragged. Sorted length stacking is rapid, and with close sticker spacing and good sticker alinement, warp is restrained. Sticker tiers are at the ends of trimmed stock and uniformly spaced in between. In stacking untrimmed sorted-length stock, one end of the package is made square at the bumper-board end of the stacking stall or by the "even ender" on the automatic stacker. The sticker tier is at or very near the end. The other end of the package is ragged to the extent that

M 134 969

Figure 71.—At this mill hardwoods are sorted for species and thickness and stacked into unit packages from a conventional green chain. Sticker guides are pitched to develop a package for placement on foundations that are sloped.

M 134 970

Figure 72.—Live rolls convey the boards to the ends of this multiple-slot sorter. The operator can sort for thickness, width, and length.

the board lengths vary from the nominal length. The sticker tier is back far enough from this end to support most of the boards.

Random-length lumber can be box-piled to produce unit packages with both ends square or with one end square and the other end ragged (fig. 79). As box-piled unit packages with a ragged end are difficult to transport without stickers being jolted out of alinement at the ragged end, the unit package should be built up with both ends square if at all possible. In box-piling random-length lumber, the long pieces are placed at the edges of the package and the shorter lengths placed in between them. By alternating or staggering the shorter boards from one end to the other, both ends of the unit package can be made square. Sticker tiers can be located so that most of the ends of the shorter boards can be supported. Bumper boards at both ends of the stacking stall can be used to produce square ends for both trimmed and untrimmed lumber. It is desirable to have long boards at the edges of each lumber layer and this may require manual replacement of boards on the layup table as the lumber moves toward the stacker.

Random-length lumber is sometimes stacked with both ends of the unit packaged ragged. Sticker tiers are placed near both ends to give support to most of the boards, yet the overhang at each end can be appreciable. Footage losses due to warp may be serious in warp-prone woods. Redwood, having low shrinkage coefficients, is often stacked into packages with both ends ragged (fig. 80); losses due to warp apparently are not enough to warrant box-piling or sorted-length stacking. End checking and end splitting can, at times, be a problem.

Package Widths and Board Spacing

The width of the stacked unit package is determined by the subsequent drying process and the load capacity of the forklift truck. Unit packages for yard drying only are likely to vary between 3 and 6 feet, with 4 feet the most common width. When the lumber is to go from the yard to the dry kiln, the unit packages vary from 3½ to 8 feet wide, depending on the width of the kiln truckloads and whether the loads are to be one or two packages wide. The packages are usually

57

4 feet wide for air drying when subsequent drying in package-loaded kilns is anticipated. Forklift trucks can handle packages up to 8 feet wide, but package height is controlled to keep the total weight within the safe loading capacity of the machine.

The boards making up a course in a unit package may be stacked with or without spaces between them. Boards in unit packages that are 4 feet or less in width are usually stacked edge to edge. There is generally one space in each course because the sum of the board widths does not equal the exact width of the package. If the packages are to be placed on kiln trucks later, or loaded into a package-loaded dry kiln, it is desirable that both sides of the package be square and plumb and that all courses have boards extending to the sides. This means that the space of the course should be within the course and not at the side. In mechanical stacking, this space is usually developed between the first two boards of a course that is laid down by the stacker arms. The stickerman or stacker operator makes sure the first board is against the sticker guide frame.

If the unit packages are wider than 4 feet and the species being stacked is relatively slow drying or inclined to blue stain, the boards on each course may be spaced. The board spacing is generally about an inch but can be increased if the package needs to be "opened" up. In hand-stacking unit packages, the board spacing is done manually but stackers acquire a skill in placing boards on the stickers as they lay up the course so that getting reasonably good board spacing is not a real problem (fig. 81). Some mechanical stackers can lay down the boards of a course with uniform spacing between boards. The equipment can be adjusted to make the board spacing larger or smaller. Wide unit packages for forklift yards are sometimes stacked with flues or chimneys.

M 134 971

Figure 73.—The boards drop out into bins alongside the multiple-slot sorter. A two-man crew stacks the lumber into unit packages.

58

Figure 74.—This two-slot edge sorter is chain driven. The boards are sorted for width and length. Packages can be stickered alongside the sorter but here the lumber is being bulk-piled for transport to a stacker.

Figure 75.—The buggies of this pocket drop sorter are transferred to a buggy dumper that feeds a mechanical stacker.

59

M 134 974

Figure 76.—The boards from the sawmill are fed onto the inclining tug chain of this dry-type sorter. They are ejected into the proper tray depending on length, thickness, width, or grade.

M 98512 F

Figure 77.—Lumber sorted for thickness and length is stacked in a stacking jig with sticker guides on one side.

Package Height

The height of the package or the number of layers in the package is determined by the thickness of the lumber being stacked, the thickness of the stickers being used, and the loading or carrying capacity of the forklift truck or carrier. A standard height is 4 feet, and most forklift trucks can transport two packages, one above the other (fig. 82). Units that are a full kiln truckload in width and height are yarded for air drying with special straddle buggies (fig. 83). Unit packages that are 6 to 8 feet wide, if high enough so that only two packages are needed to load a dry-kiln truck, generally require lift-truck equipment designed to carry the heavier loads.

Pile Widths and Heights

Pile widths in a forklift yard are determined by the width of the unit package. Sometimes the air-drying pile of unit packages of softwoods is made two units wide with a relatively small space between the two tiers on the pile foundation. Pile height in a forklift yard depends upon the equipment used for placing unit packages in the pile. Piles of unit packages built up with forklift trucks are generally not over about 20 feet high and consist of four packages (fig. 84). When the pile of packages is much higher than 20 feet, it may be necessary to handle the two upper packages together for placement.

60

M 115 737

Figure 78.—Sorted-length boards of the same thickness are being stacked by machine. A course of lumber is made up on the layup table and carried to the package by stacker arms and laid down. The stickerman places stickers in the guides.

With piles of packages that are relatively narrow, a high pile may not be stable and may tip (fig. 85). This is particularly true when packages are poorly made—not box-piled—or there are displaced boards or stickers in the lower packages. Tipping of the piles may partially close spaces between the piles and retard air movement, thus affecting drying, or present difficulties when the piles are to be dismantled. Piles that fall are a danger to workmen and could damage machines. High piles, however, do offer some advantages. More lumber can be piled in a given yard area, and more lumber can be protected by a given-size pile roof.

Use of Bolsters

Bolsters separate the unit packages (fig. 86) to permit ready exit and entrance of the lift-truck forks. The bolsters are usually 4- by 4-inch wood members and as long as the stickers. Bolster placement is critical as "out of line" bolsters between packages can cause warp (fig. 87). The bolsters should be placed immediately above the tier of stickers of the bottom package in the pile. The

M 121 573

Figure 79.—In this stacking jig designed for a one-man crew, one end of the package is made square by bumping the boards against the end board. The other end is ragged.

61

Figure 80.—Random stacking of packages is tolerated in redwood as warp of the unsupported ends is not serious.

packages on top should also be lined up so that the sticker tiers and all bolsters between packages are in vertical alinement. Thus, distortion or warp is restrained. Bolsters long enough to span two unit-package piles are often used between the top two packages in the pile to create greater stability.

A "side shifter" carriage on the forks of the lift truck aids in placing the packages so stickers are

Figure 81.—This package of random-width hardwood boards is being box-piled with a space left between each board.

Figure 82.—The standard height of unit packages is about 4 feet and most forklift trucks can maneuver with two packages.

63

Figure 83.—These large stickered unit packages of hardwoods are transported to the air-drying yard by a special carrier

Figure 84.—Sorted-length, stickered packages placed four high in a line-type forklift yard.

64

Figure 85.—Narrow packages may tip in spite of a good foundation. A bolster tie between the top packages in adjacent piles may be a simple answer.

alined (fig. 88). Bolsters still must be carefully placed, and the lift-truck operator must avoid jolting the bolsters out of place as he moves the upper packages in the pile into position. The bolsters are placed on the packages when the package is at ground level. Although bolsters between packages at every sticker tier would produce maxi-mum influence of superimposed loading to restrain warp, it is not considered practical, and three to five bolsters are usually used on 16-foot lumber.

For bolsters, both softwood and hardwood species are used. Because they are generally used over and over again, they should be of hardwood of the more durable species being processed at the

Figure 86.—Wood bolsters are used at every sticker tier between the package and the foundation and every other sticker tier between the upper packages in the pile. The bolster space provides easy pickup for the forklift truck.

Figure 87.—Bowed lumber results when the bolsters are not lined up with the tier of stickers. Sticker misalinement in stacking this unit package is also evident.

sawmill. Bolsters are generally rough sawn and are dried the first time out. When not in use, bolsters should be stored under cover and kept dry. They are not usually treated to prevent deterioration due to stain, decay, and insects, yet should not infect or infest the green lumber on which they will be used.

Piling Modifications

The yarding arrangement of main alleys, cross alleys, rows, lines, and pile foundations within a row or line is generally pretty well fixed. In completely paved yards with portable foundations, however, areas or blocks can be rearranged to either hasten or retard the drying rate. Even in line-type forklift yards having fixed foundations, lumber might be piled in every other line when drying conditions are poor. An alternate practice would be to stagger the piles between the two

66

M 134 985

Figure 88.—In placing this package on the pile, the fork-lift operator can move the package sideways with the "side shifter" to aline the sticker tiers with the lower packages. Note that the bolsters are in place on the lower package.

foundations in the line. Yards with long rows between main alleys may have a foundation system that will permit changing pile spacing, increasing the spacing when drying weather is expected to be poor and reducing the pile spacing when the anticipated drying potential is good.

Pile heights can be changed from six packages high to one or two packages high to reduce the volume of lumber exposed to severe drying conditions. This is seldom done, however, for during these periods of good drying weather the sawmill production is apt to be at its peak and the yard-holding capacity per acre must be fully utilized.

In the arid and semiarid regions of the West and Southwest, softwood lumber may be air-dried in self-stickered piles (fig. 89). These hand-built piles are square and level or with slight slope but no pitch. Some of the boards being piled are used for stickers or crossers to separate the layers of lumber. These stock stickers are sold along with the rest of the lumber. The number of stickers per layer varies from three on the shorter boards to five on 16-foot stock. The boards may or may not be spaced, but it is not uncommon to see boards that are 8 inches and wider spaced 1 to 2 inches apart. The piles built up of 16-foot stock will be 16 feet wide and will usually be constructed with a single chimney in the middle of the pile.

In these wide hand-built piles, the importance of chimneys and flues, as well as proper board spacing, is evident for good drying.

The makeup of the unit package itself is usually fixed as to size and method of course layment. But edge-to-edge-stacked packages can be changed to spaced board stacking. Also the spaces between boards can be increased or decreased as air-drying conditions warrant.

Stacking Lumber in Hand-Built Piles

Although much lumber is air dried in unit packages, piles are still hand built in some yards. Hand-built piles may be erected on sloped foundations. As the sloped pile is built up layer by layer, the board ends at the front end of the pile are placed to slightly overhang the course beneath and create a general forward pitch to the whole pile. The sloped and pitched, hand-built pile uses stickers that are as long as the pile is wide. Such stickers are used on both hardwood and softwood lumber and are usually nominal 1-inch material 6 to 8 feet long, depending upon pile width. In piling thick softwoods they may be 2- by 4-inch stickers and as long as 16 feet.

Hand-built piles may be up to 30 feet in height. Elevated tramways provide one way whereby these heights can be attained. Portable lumber elevators also permit piles to be built this high from roadways or ground-level tramways. Hand-built piles stacked without an elevator seldom are higher than 16 feet.

Board Spacing

Vertical flues or chimneys are built into the wider handbuilt piles by spacing groups of boards. When these openings in the pile are less than 6 inches wide, they are arbitrarily called flues; when greater, they are called chimneys. Chimneys or flues or both are built into the wide, hand-built air-drying pile to provide passageways for the movement of air. The downward movement of air in these vertical openings induces the entrance of warmer and drier air at the top and at the sides of the lumber pile.

Where flues are used, they are narrower and in greater number than where chimneys are used.

Figure 89.—This softwood lumber is self-stickered for air drying in a hand-built pile. The boards in each layer are spaced and a chimney was built into the pile.

Flues are usually used with single-width stock or with stock of narrow widths. Chimneys are generally used with random-width stock. A common ratio of flue or chimney width to pile width is about 1 to 5.

Stacking of Random-Length Lumber

Sloped and pitched, hand-built piles of lumber are usually constructed of lumber sorted for length. If random-length lumber is stacked, box piles are made with the boards all flush with the front or pitched end of the pile. The other end of the pile is ragged and good sticker support for the various lengths is lacking. Lumber is sometimes grouped according to length; for example, 6, 10, and 12 feet and 8, 14, and 16 feet. Since the shortest boards are half as long as the longest boards, the scheme aids stacking. The ragged end is better supported by a sticker tier.

Level, hand-built piles can be box-piled with both ends square to incorporate good sticker restraint to warp. Grouping 6-, 10-, and 12-foot lengths and 8-, 14-, and 16-foot lengths into separate piles will result in less void space and better stickering. Level, hand-built piles are usually provided with pile roofs to keep rain entry to a minimum.

Crib Piling

Crib piling is a method of stacking lumber for air drying without stickers. Separate piles are made for each length of lumber by cribbing three tiers to form a triangle. The tiers may be single, double, or triple (fig. 90). The first plank rests on a support that may be a block of wood placed upon the ground at each end. One end of the second plank crosses the first plank above a support, and the other end rests upon a third support of the

M 134 987

Figure 90.—This western softwood was sorted for width and crib-piled. Shorter length boards were placed on the inside of the crib.

triangle. The third plank closes the triangle. In succeeding courses, this crib work is carried to a height convenient for one-man stacking. The ends of all of the planks cross each other directly along the supports.

Crib piling is sometimes used by small mills cutting softwoods because the piling can be done by one man, but it requires considerable yard space (fig. 91). The method encourages rapid drying, except where the boards cross each other. A modification to improve the method is shown in figure 92. The boards in each pile preferably should be of the same length. If mixed lengths are piled, the triangle is made to fit the length of the shortest boards; the ends of the long boards project beyond the corners. The ends of the boards that contact each other are likely to stain because of the retarded drying—unless the lumber is dipped in an antistain chemical—while the long, unsupported midparts of the boards are likely to sag and warp. Crib piles are not provided with roofs. The exposure of the boards to the elements is likely to cause checking and warping.

Special Piling Methods

The unit packages piled for air drying or the hand-built piles are usually considered flat piles. Other methods that have more or less limited use are end piling and end racking. End piling is approximately equivalent to tilting a flat pile on end until the boards are nearly vertical (figs. 93, 94). In end racking, the boards are placed upright and cross each other to form an "X" or an inverted "V" (fig. 95). In both end piling and end racking

the rows of boards are supported by frameworks. Neither of these types of pile is roofed.

With both end piling and end racking. it is important to support the lower ends of the boards above the ground surface. The lower ends of end-piled boards may rest on a spaced-board platform or on boards, planks, edging strips, or other low-grade lumber laid directly on the ground. In end racking, the lower ends of the boards usually rest on timbers or planks placed directly on the ground.

Both of these piling methods have disadvantages and advantages over flat piling. End piling and end racking can be done by one man, while flat piling methods generally require at least a two-man crew. End piling and end racking promote rapid surface drying and consequently reduce the likelihood of staining. But treating lumber with a fungicide is common today, so this piling method is rarely used as the primary method of stain control. End-piled and end-racked lumber are more susceptible to surface and end checking and warp than flat-piled lumber. The boards are often wetter at the bottom than at the top ends.

Stickering Lumber for Air Drying

Stickers perform several important functions in air drying. They separate the courses or layers of lumber, permitting air to circulate over the broad faces of the boards (fig. 96). Each sticker carries its proportionate share of the weight of the pile above it. Stickers, in combination with the loads that are imposed on them, tend to keep the boards flat and thus restrain warp. Finally, the position of the stickers at the two ends of a pile may reduce end checking and splitting, and warping near the board ends.

In carrying out these roles, stickers must contact wet boards—and the contact area is somewhat retarded from drying (fig. 97).

M 100 832

Figure 91.—A crib-piled yard of southern pine.

69

Figure 92.—In these crib piles of southern pine lumber, two of the boards are lapped to reduce the contact area at the crossing points. The piles are low to reduce handling.

Species of Wood for Sticker Manufacture

Stickers are often made from lumber that is available at the plant where the air drying is performed, but sometimes they are purchased. In any case, stickers should be straight grained. durable, and strong so they can be used over and over again. The oaks, beech, and hickory make good stickers for hardwood operations. Douglas-fir and larch are used for stickers in the Pacific Northwest. Southern pine and cypress heartwood are used on softwoods in the South. It is preferable to make stickers from heartwood rather than sapwood. Stickers made of sapwood may become in-fected with stain or decay fungi and thus infect freshly cut lumber on which they are laid. Green edgings from lumber have been used for stickers but do not perform as well as stickers made from thoroughly air-dried or kiln-dried lumber.

Sticker Size

For piling hardwoods, the stickers are usually made from 1-inch lumber, either rough or dressed to about $25\!/\!32$ inch, and from 1 to 2 inches wide, usually $1\frac{1}{4}$ inches wide. For piling softwoods, the 1-inch-thick stickers are generally wider, as much as 4 inches. Sometimes the narrower boards of

70

Figure 93.—Sorted-length lumber end-piled on a wood-platform.

the lumber being stacked are used as stock stickers.

Sticker length is determined by the width of the unit package or pile being constructed. Four-foot packages being stacked by hand in a framework of sticker guides or on an automatic stacker will use stickers about 50 inches long. Hand-built piles in the yard will use stickers that are 6 feet long on a 6-foot pile. The pile is usually a few inches narrower than 6 feet so that the stickers project enough for the stacker to see the sticker tier location as he constructs the pile.

All stickers in the same layer should be of uniform thickness to minimize warping (fig. 98). Miscut lumber from which stickers are being ripped should be surfaced one or both sides prior to ripping in order to produce stickers of uniform thickness.

Rough Versus Dressed Stickers

Whether the sticker surface contacting the lumber is rough or dressed probably does not greatly influence the drying rate of the wood under the sticker, although some yard supervisors prefer to use stickers with rough-sawed surfaces. On the other hand, the operation of automatic stackers seems to be less troublesome when stickers made from surfaced lumber are used.

Sticker Alinement

Stickers should be in good vertical alinement in both unit packages and hand-built piles (fig. 99). If the stickers are not alined, the superimposed weight will not be transmitted from one to another by direct compression, but will act on the span of the boards to cause warp (fig. 100). Sticker guides are more essential for stacking unit packages of lumber than for hand-built piles. Guides not only assure good sticker alinement within a package, but they also regulate the sticker spacing so that it will be the same from package to package. Uniform spacing is necessary if good alinement of stickers, bolsters, and foundation cross beams is to be obtained in a pile consisting of several packages. In stacking hand-built piles, sticker guides are seldom used and sticker alinement depends on the care and skill of the lumber stacker. In building up sloped and pitched piles, the sticker tiers should follow the pitch of the pile.

Sticker Spacing

The space between stickers in a course of lumber depends on the thickness of the lumber. Nominal 1-inch lumber requires closer sticker spacing to control warp than 2-inch material. Stickers retard drying to some extent at the areas where they contact the boards, so the fewest number of stickers that will produce the desired warp control is best. Fewer stickers are used with softwoods than with hardwoods, so softwood stickers are wider to provide sufficient bearing surface and avoid crushing the sticker or the lumber. Five stickers are generally used on 1-inch-thick, 16-foot softwoods. From five to nine stickers are used on 1-inch, 16-foot hardwood lumber (fig. 101). Spacing stickers 16 inches to 2 feet apart is common for hardwood lumber, particularly for unit packages where sticker guides keep them in place when the courses of lumber are laid down.

Stock Stickers

Hand-built piles that are self stickered require three to five stickers for a pile length of 16 feet. If random-width boards are being piled, the narrower ones will be selected as stickers.

Special Sticker Designs

Stickers used in stacking lumber for air-drying are usually rectangular in shape. The broad faces of the sticker contact the stacked lumber. Where sticker marking is associated with chemical brown stain, grooved stickers have successfully lessened

71

M 134 989

Figure 94.—Mixed-length lumber end-piled on boards laid on the ground.

the problem (fig. 102). Here 2-inch-wide by ⅞-inch-thick hardwood stickers have a groove approximately 1⅜ inches wide machined out on the two sides to provide 5/16-inch bearing areas at each edge of the sticker. The grooved sticker presumably allows some air circulation under the stickers, reducing the likelihood of chemical reaction causing the stain.

Sticker Requirements for Stacking

The automatic stacker ejects stickers from the bottom of the sticker magazine. If the stickers are not straight and uniform in cross-sectional dimension and length, the stacker will not function properly. Warped and end-battered stickers create troubles in stacker operation so the stickermen must reject poor stickers.

In semiautomatic stacking, the stickers are laid on the layers of lumber by hand. Sticker guides on both sides of the package are usually provided. Sometimes the sticker guide slots are considerably wider than necessary, resulting in some offset stickers in the tiers. As the stickers are hand laid, some sticker warpage can be tolerated, but badly crooked stickers should be discarded. Badly bowed stickers may require quite a few lumber layers on top to flatten them out. Twisted stickers should be rejected because they cause the same problems as do excessively thick stickers mixed in with others of uniform thickness.

For hand laying, slightly shorter stickers can be used. The stickerman can use the sticker guides opposite him for one end and place the other end in line with the sticker guide slot on the near side. Where stacking operations are fast, however, sticker misalinement on the near side can result.

72

Figure 95.—Sorted-length, random-width hardwood lumber end-racked to obtain rapid surface drying.

M 134 990

Figure 96.—In order for air to circulate over the broad surfaces of the lumber, the layers or courses of lumber are separated with wood stickers.

M 134 991

Figure 97.—The area under the sticker is retarded from drying and stain sometimes results. This hardwood board was stained at the sticker location.

Handling and Storage of Stickers

Stickers used in air drying lumber last longer than those used in kiln drying. They are not subjected to repeated high temperatures and they are not handled so frequently. Sticker life, however, can be shortened by rough handling. Stickers that have dropped out of the unit package may be run over by transport machines and broken.

73

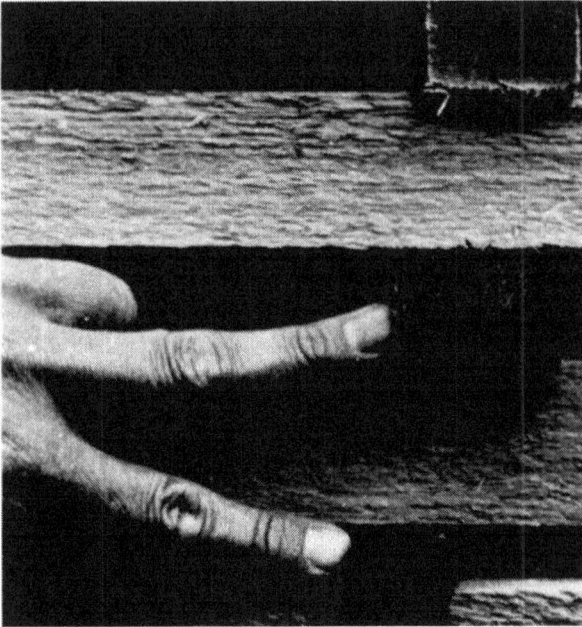

Figure 98.—Thick and thin stickers in the same sticker space cause warp. If sticker thickness varies, stackers seldom have time to select stickers of the same thickness for the course.

Figure 100.—Lumber warps when the stickers are not in good alinement and the bolster is offset from the sticker tier.

Figure 99.—This hand-built pile of softwood lumber illustrates excellent sticker alinement.

Figure 101.—This 16-foot southern hardwood lumber is stacked with stickers every 2 feet. Stickers are in good alinement, foundation cross beams are located at every sticker tier, and bolsters support every other sticker tier.

Figure 102.—These hardwood stickers are grooved to reduce the contact area with the lumber. The possibility for sticker marking in hard maple lumber at the sticker areas is lessened.

Figure 103.—These stickers in the rear alley in a hand-built pile yard interfere with air movement in the yard and are exposed to conditions resulting in infection by stain and decay fungi.

Stickers used in hand-built pile yards remain in the yard and, unless they are stored in special shelters, are exposed to wetting by rains. When hand-built piles are taken down, the stickers are often thrown on the ground between the piles or in the rear alley (fig. 103). If they rest directly on the ground, they are exposed not only to rain, but to ground moisture as well. They are sometimes placed on extended cross beams of the pile foundation, but even here are partially exposed to the rain.

Wetting may cause the moisture content of the stickers to increase to the point where stain or decay fungi can grow. This may shorten the life of the stickers, and fresh lumber piled with them may become infected. Poor sticker storage and handling and exposure to wetting and redrying may also cause the stickers to warp, which will contribute to their breakage when they are used.

Forklift yards usually develop a sticker-handling system. The unit packages are generally built at some central location. This requires that the stickers be returned from where the packages are taken down to the place where the stacking is done. If the packages of lumber are kiln dried

after air drying, the stickers are returned to the stacker after breakdown of the kiln-dried package in the planer mill, the rough mill in a factory, or at the boxcar door. At the package-breakdown location, the stickers are often loaded into racks or bins that can be carried to the stacker by the lift truck or carrier (fig. 104). In a forklift yard, it

Figure 104.—The handling of stickers is facilitated if they are racked for transport by carrier or forklift truck.

75

is possible to provide protection for the stickers from the weather by keeping the carrier racks or bins under cover.

Sticker Salvage

Stickers represent a considerable investment. Although the cost of each individual sticker is not great, the piles on the yard contain a considerable number of them. To what extent stickers are salvaged depends on their value. Today some plants salvage broken stickers by trimming long stickers to the shorter lengths used in unit-package operation. Where only short or long stickers are used, the broken stickers with some salvage value are trimmed, scarfed or finger jointed, glued, and cut to length. Used stickers have sale value when no longer needed.

For Additional Information

Blew, J. O.
    1959. Symposium on method of conditioning wood prior to preservative treatment. Proc. Amer. Wood-Preserv. Assoc.
Connor, R. M.
    1952. Packaging and handling of lumber, dimension and hardwood flooring at a hardwood sawmill, furniture factory and flooring plant. J. Forest Prod. Res. Soc. 2(5) : 28–31.

Fullaway, S. V., Jr., Johnson, H. M., and Hill, C. L.
    1928. The air seasoning of western softwood lumber. U.S. Dept. Agr. Bull. 1425.
Glesne, N.
    1959. Small sawmill lumber handling. Forest Prod. J. 9(7) : 9A–10A.
Hyler, J. E.
    1959. Modern sawmilling. South. Lumberman 198(2473) : 33–36 (April 15); 198 (2474) : 37–41 (May 1).
McLaughlin, T. P.
    1955. What about lumber stackers? South. Lumberman 191(2388) : 72–74 (Oct. 1); 191(2389) : 63–64 (Oct. 15); 191(2390) : 60–62 (Nov. 1); 191 (2391) : 46 (Nov. 15); 191(2392) : 44 (Dec. 1).
Rice, W. W.
    1959. Use stacking jig for better lumber packages. USDA, Forest Serv., Central States Forest Exp. Sta. Note 131, St. Paul, Minn.

    1964. Learn profitable lumber stacking. Hitchcock's Woodworking Dig. 66 (2) : 44–46.
Smith, H. H.
    1954. Seasoning of California hardwoods. USDA, Forest Serv., Calif. Forest Exp. Sta. Tech. Pap. No. 5, Berkeley, Calif.
U.S. Department of Commerce
    1929. Seasoning, handling, and care of lumber (Manufacturer's Ed.) Nat. Com. Wood. Util. Rep. No. 9.

CHAPTER 6

AIR-DRYING DEFECTS—CAUSES AND REMEDIES

Losses in value and footage resulting from defects that develop in lumber during air drying increase the cost of air drying. If known, such losses are directly assessed against the air-drying process. Air-drying defects may be caused by shrinkage, by fungal infection, by chemical action, or by insect infestation. Shrinkage causes end checks and end splitting, surface checks, honeycomb, and warp. Exposure of lumber directly to weathering conditions aggravates these shrinkage defects, and extended yarding after the lumber is air dry accelerates the rate of grade deterioration or footage losses. Fungus infection causes blue stain in the sapwood as well as decay and mold. Chemical reactions cause chemical brown stain, and sticker marking is one form of this discoloration. Insect infestation results in pith flecks, pinholes, and grub holes.

Drying defects that do not degrade rough lumber may cause footage losses during machining. Warp may cause skips or splits in the dressing operation. Knots that are loosened in softwoods during drying may be knocked out during planing and knots that are checked may be broken.

Defects Resulting From Chemical Change

Certain stains or discolorations develop in lumber during air drying in addition to those caused by fungi or by general weathering. These stains result from chemical changes that occur in the wood. Called chemical brown stains, they darken the wood to colors ranging from buff to dark brown. The commercial species that are subject to objectionable chemical stains are ponderosa pine, the true white pines, western hemlock, noble fir, redwood, and several hardwoods including ash, maple, birch, hickory, and magnolia.

In pine, chemical brown stain develops in either sapwood or heartwood. It is more prevalent in lumber sawed from logs that have been cut for some time than in lumber from newly cut logs. It is also more prevalent in boards that have been solid piled for 2 or more days immediately after sawing than in boards stacked for drying immediately after sawing. The stain develops during hot, humid months and usually is observed on the surfaces of the boards, but it may penetrate deeply

into the wood. In some cases, it occurs inside a board and does not show on the surface. The stain results from a concentration of extractives that are transported by the water and deposited at the point where the water is vaporized or absorbed as bound water. These extractives are believed to be sugars and amino acids, which are present within the free water or are formed through enzymatic action immediately after felling or during solid piling of the lumber. The enzymic action can be slowed down by dipping the green lumber in enzyme-inhibiting chemicals.

Sticker marking, a form of chemical brown stain, develops in woods such as maple, white ash, and magnolia during the warm, humid summer months. The discoloration, which often does not surface out of the rough, dry boards, can be significantly reduced by using dry stickers and subjecting the lumber to good drying conditions immediately after stacking. Narrow, grooved, or toothed stickers are sometimes used to reduce the contact area between the sticker and the lumber and thereby keep sticker marking to a minimum. Shed-fan air drying and forced-air drying, or low-temperature kiln drying, are very effective in preventing sticker marking in these woods during the warmer months of the year.

Chemical brown stain can be reduced by conditions that encourage rapid drying. Rapid air drying can be promoted by keeping the yard surface free from vegetation and debris, using high and open pile foundations, increasing the spacing between piles in the rows, and opening the unit packages by increasing board spacing and constructing more chimneys in the hand-built piles.

Defects Resulting From Fungal Infection

Stains

Stains in wood are caused by fungi, which are microscopic plants that grow in the wood and use parts of it for food. Stains are confined largely to sapwood of both softwoods and hardwoods and are of various colors. The so-called "blue" stains, which vary from bluish to bluish-black, are most common. Except for toughness, blue stain has little effect on the strength properties of wood, although it does cause degrade where color is important.

Blue stain is likely to develop where air drying

is retarded. It is most likely to occur during the warm, damp seasons of the year. It occurs in flat piles of self-stickered lumber, where green boards are used for stickers, and in end racking and crib piling, where the boards themselves come in contact. The likelihood of blue staining can be reduced by using narrow, dry stickers and by opening up the yard and the piles to encourage rapid air drying.

Fungal growth can be prevented in lumber by quickly drying it to a moisture content of 20 percent or less and keeping it dry. As air-drying conditions may not always be favorable to prevent the growth of staining fungi, chemical treatment of the freshly cut lumber may be necessary. The chemical treatment is usually accomplished by dipping the lumber or spraying it with a suitable fungicide [1] (fig. 105). If, however, the lumber has already become infected, the fungus may have penetrated so far below the surface that the organism is not completely killed by dipping. When such lumber is air dried slowly, the interior portions may blue stain, although the surfaces may remain bright. The effectiveness of the antistain chemicals may also be reduced by leaching, if the

lumber is exposed to wetting while in the yard.

Successful stain control with chemicals not only depends upon immediate and adequate treatment but also upon proper handling of the lumber in the air-drying yard. The yard and lumber storage areas must be kept as sanitary as possible to reduce chances of infection. The fungus is propagated by spores which are produced on the surfaces of the wood when the fungus has reached a certain stage. These spores are airborne, and are practically always present. They infect freshly sawed lumber by coming to rest on the surfaces; if the conditions of air, moisture, and temperature are favorable, they develop quickly into the fungus. Sap-stain fungi can grow at temperatures of 35° to 100° F. Spores, although generally carried by the wind, are also carried by insects, and where the insect burrows into the sapwood, the spores are carried into the burrows.

Mold

Mold is also propagated by airborne spores. During warm, moist weather mold grows on wood surfaces and it also penetrates the wood. As the hyphae or threads are colorless, they do not stain the wood. The discolorations of wood surfaces are caused by the fruiting bodies. Under exceptional conditions, mold may develop to a point where it

[1] Check with your County Agricultural Agent or State Agricultural Experiment Station for approved recommendations.

M 38233 F

Figure 105.—These dip-tank arrangements (A, B, and C) and a spray chamber (D) are designed to treat green lumber with a fungicide.

restricts air circulation in certain portions of a pile, and thereby retards drying. The measures used to reduce or control blue stain apply also to mold.

Decay

Decay or rot is caused by fungi that not only discolor wood but actually destroy it. Decay, blue stain, and mold organisms all thrive under similar conditions of moisture content, air, and temperature but decay requires somewhat longer to develop. Freshly sawed lumber may be infected by airborne spores or by contact with decayed foundation timbers or stickers (fig. 106). The best way to combat decay is to dry the lumber to a moisture content of 20 percent or less as rapidly as possible. In some cases, it may be necessary to treat with a suitable fungicide.[2]

Decay is frequently present in the living tree and lumber sawed from the logs will contain the organisms. Some of these decay fungi may continue to develop in the lumber as it dries.

Defects Resulting From Insect Infestation

Wood in all stages of drying, from the green condition to completely dry, may be subject to attack by insects. Piles of lumber in an air-drying yard are sometimes infested. Debris, in the form of broken timbers or stickers, provides breeding places for insects that may spread to the lumber.

Spraying the logs with a suitable insecticide will control the insect.[2] The addition of one of the fungicides previously mentioned for controlling stain, mold, and decay will keep the lumber bright.

Powder-post beetles attack both hardwoods and softwoods, both freshly cut and air-dried lumber. The sapwood of hickory, ash, and oak are particularly susceptible to attack. Damage is indicated by holes left in the surface of the lumber by the winged adults as they emerge and by the fine powder that may fall from the wood. Sterilization of green lumber in saturated steam at 130° F. or at lower relative humidities at 180° F. for 2 hours is effective for 1-inch lumber. Thicker lumber requires a longer time. As heat-sterilized wood will

[2] Check with your County Agricultural Agent or State Agricultural Experiment Station for approved recommendations.

not prevent subsequent infestation, good yard sanitation is essential to check infestation by these insects.

Defects Caused by Shrinkage

When a log is sawed into lumber the drying process starts, and soon after stacking in the air-drying yard, shrinkage of the board begins. The stresses set up in the surface zones of the lumber by shrinkage may cause deformation or failure. Because the amount of shrinkage varies with the species of wood and the grain patterns of the lumber, a change in shape usually results. If the stresses exceed the strength of the wood, failures may develop, such as various types of checks, splits, and cracks.

Checks

Checks are failures of the wood that develop along the grain because of drying stresses. They are of three sorts—end, surface, and honeycomb. Some woods are inclined to check more readily than others. A very general classification of the tendency to check of some species is:

Tendency to Check

Low	Intermediate	High
	SOFTWOODS	
Baldcypress	Firs, true	Douglas-fir
Cedar	Hemlocks	Larch,
Pine, sugar	Pine, jack	western
Pine, loblolly	Pine, lodgepole	
Pine, shortleaf	Pine, longleaf	
Redwood	Pine, ponderosa	
Spruce	Pine, red	
	Pine, slash	
	Pine, white	
	HARDWOODS	
Alder	Ash	Beech
Aspen	Birch, yellow	Oaks
Basswood	Butternut	Sycamore
Birch, paper, and	Elm, rock	Tanoak
and sweet		
Cherry	Hackberry	
Cottonwood	Hickory	
Elm, American	Maple, sugar, and	
	bigleaf	
Magnolia, southern	Pecan	
Maple, red, and	Sweetgum	
silver		
Tupelo	Walnut	
Yellow-poplar	Willow	

End checks originate on end-grain surfaces, and appear as radiating lines pointing toward the pith or heart center of the tree (fig. 107). They occur at the junction of the wood rays and the remainder of the wood cells, or within the wood rays. Once started, they become wider and, by extending radially and longitudinally, develop into splits. Surface checks are similar separations of the wood

Figure 106.—These decayed pile foundation members can infect green lumber.

under stress, but they occur on tangential or flat-grain faces (fig. 108). They become longer by extending in the longitudinal direction of the grain of the wood and deeper by extending in the radial direction.

Hot, dry weather immediately after piling is likely to cause checking. End checks will probably develop first, followed by surface checks. End checks and surface checks will probably be more severe in those parts of the pile that are more fully exposed—that is, the ends, sides, and top parts of the pile. Checks will be particularly severe in the upper surfaces of the boards in the top course if the pile is not roofed.

Footage losses due to end checking and end splitting can be serious, particularly in lumber 1¼ inches thick or more and in the higher value grades.

End coating the lumber immediately after trimming to length retards end drying, which causes these failures. A number of proprietary end coatings are available. The emulsified waxes that can be applied by either brushing or spraying are effective for use on lumber being air dried.

End checking frequently begins on the log yard. If high-value logs are to be left in open storage for long periods of time, it may pay to end coat both ends with emulsified wax to minimize end checking. An alternate method would be to sprinkle the dry log deck with water.

Honeycomb

Though not so common in air drying as end and surface checks, honeycomb or internal checking may occur (figs. 109, 110). Honeycomb failures may result from surface or end checks that have closed on the surface, or they may result from tensile failures entirely within the interior of the piece. Such end checks or surface checks may not be serious in themselves, but they can penetrate and extend longitudinally into the board. Sometimes the presence of honeycomb is indicated by surface depressions or grooves, but usually it cannot be detected until a piece is dressed or sawed. Remedies for preventing end and surface checking may also be used to prevent honeycomb.

Shakes

Failures that do not conform with the usual definition of checks sometimes occur under shrinkage stresses. These failures are caused by radial shrinkage stresses. They are checklike openings located within or at the junction of the annual

Figure 107.—End checks in thick oak lumber. The end checks can extend to develop splits.

80

M 101 115

Figure 108.—These oak boards surface checked in the early stages of air drying. These surface checks can penetrate deeper into the wood, and when they close at the surface, the internal failure appears to be honeycomb.

rings. They may originate on end-grain surfaces and look like end checks, except that they follow the rings rather than the rays. These ring failures are not considered a drying defect unless they develop into splits, or internal honeycomblike openings (fig. 111).

Splits

Splits are longitudinal and radial separations of the wood. Usually they occur radially. They are generally located at a board end, but occasionally occur along the length of a board, particularly near where it is crossed by a sticker. A split along the length of the board may or may not extend completely through the thickness of the piece.

As mentioned, splits generally start as end or surface checks. Splits are sometimes associated with longitudinal stresses that were in the log and in the board when it was freshly sawed. When a split originates, the longitudinal stresses cause it to open wide and to extend along the length of the piece. The length of splits may be increased by rough handling or planing after the drying proc-

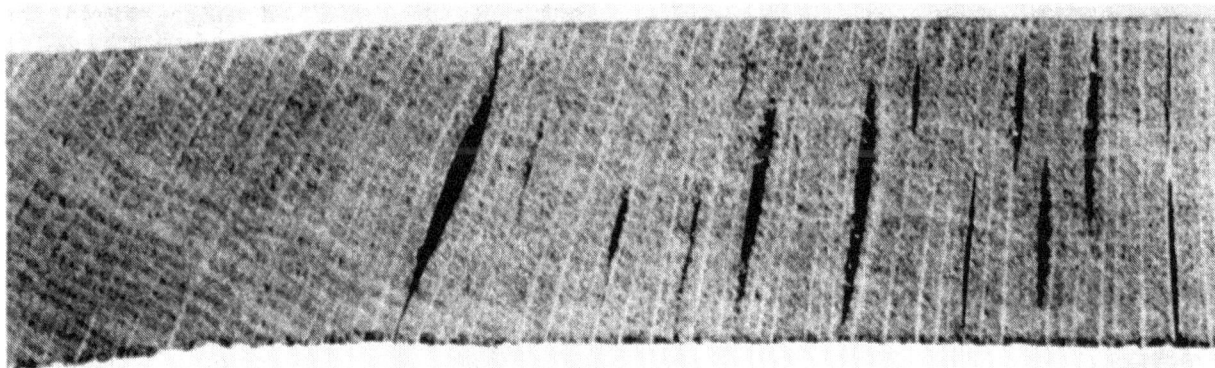

M 11293 F

Figure 109.—This cross section of an oak board is honeycombed. Most of the internal failures were caused by penetrated surface checks.

414-042 O - 71 - 5

Figure 110.—These pieces are badly honeycombed. Surface checking in air drying probably accounted for most of the damage.

or end checks, but their characteristically large width is caused by the difference between tangential and radial shrinkage. Cracks are common in poles, posts, and boxed-heart timbers. On a transverse section, a crack appears as a wedge-shaped opening extending from a face to the pith.

Cracks or checks in the heart center of the tree are not always caused by shrinkage, but are often present in the tree and in the log. They are the result of stresses to which the living tree was exposed, generated possibly by winds.

Cross Breaks

Boards containing abnormal wood (such as compression wood) in the central portions with normal wood on the edges may, when dry, contain checks crosswise of the board (fig. 11). These result mainly because the abnormal wood shrinks more longitudinally than the adjacent normal wood. The restraint by the normal wood results in tension stresses parallel to the grain in the abnormal wood; finally these stresses exceed its strength and cause failure. Cross breaks in normal wood are sometimes observed and are generally identified as compression failures that may have developed in felling the tree.

Collapse

Some woods with a very high green moisture content tend to take on a corrugated or washboard appearance when dry. The excessive shrinkage is collapse (fig. 112). It is observed more in quartersawed lumber than in plainsawed lumber. Normal shrinkage of wood causes but little change in the size of the cell cavity, whereas when wood is collapsed, the cell cavities are much smaller and often appear "caved in." Collapse is aggravated by kiln drying. Air drying green lumber cut from butt or sinker logs of redwood and western redcedar and the heartwood of sweetgum and swamp oaks reduces the amount of collapse as compared with what would be found if the material were kiln dried. Collapsed eucalypts that were air-dried or predried are reconditioned with steam to restore normality prior to kiln drying the lumber to lower moisture content values. This process has not been adapted for American woods, probably because the volume of lumber involved does not justify the added costs.

Casehardening

In air drying green lumber, the fibers in the outer zones of a board dry below the fiber saturation point, but their shrinkage is restrained by the

ess. Cupped boards containing an end split may split from end to end under the pressure of the feed rolls, or the pounding of the knives of the machine.

Cracks

Cracks have the appearance of surface checks or splits but are formed differently. Cracks occur in pieces containing the pith or heart center of the tree. They generally develop from surface checks

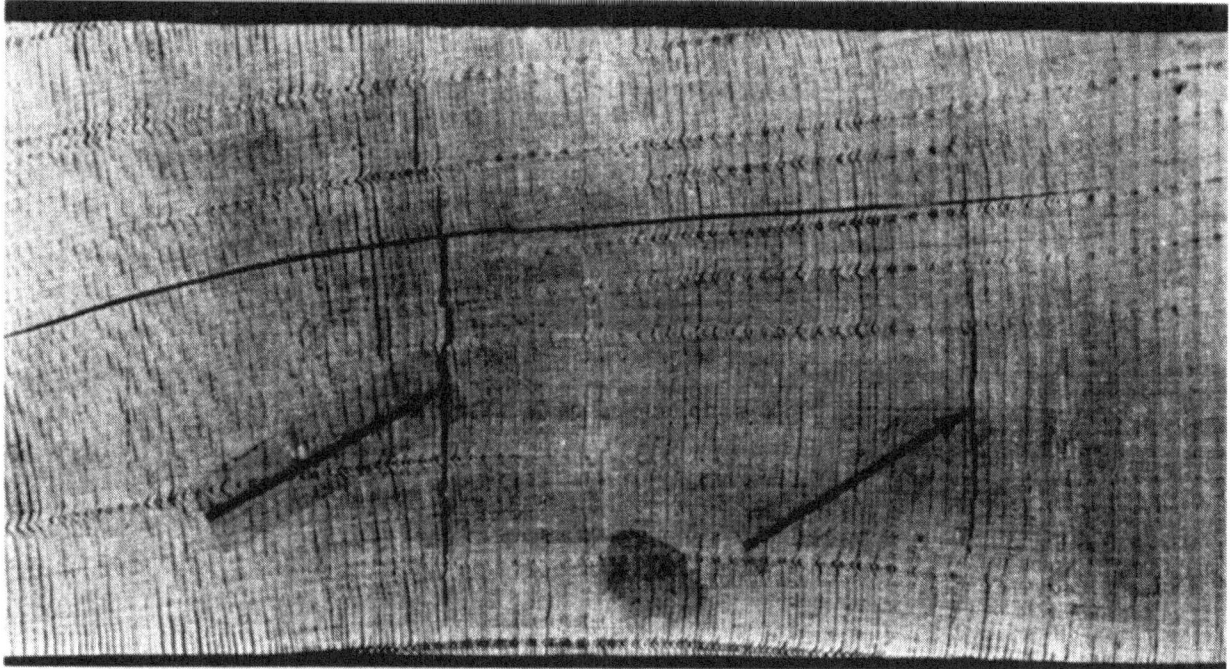

M 52591 F

Figure 111.—The arrows indicate ring separations or shake. The long horizontal failure is honeycomb.

core. A tension set develops in the outer zones. Some compression set may develop in the core. Later when the board is uniformly dry through its cross section, the drying stresses have reversed: The outer zones are in compression, and the inner fibers are in tension. This is a condition known as "casehardening." It is the normal behavior of lumber to undergo stresses as it dries and to develop set. In air drying, a considerable amount of tension set can be developed in the outer zones of a board. This tension set occurs when the board is flat, and as drying proceeds it restrains plainsawed boards from cupping. In air drying, the development of casehardening is normal and is not considered a defect.

Loosening of Knots

Knots that are not intergrown with surrounding wood generally loosen during air drying. Both the shrinkage of the wood surrounding the knots and the shrinkage of the knot itself contribute to the loosening. A knot does not shrink as much lengthwise as the board shrinks in thickness, and as a result, it tends to project above the surfaces of the dried board. Encased knots are surrounded by bark or pitch that may become hard and brittle when dry, loosening the bond between the knot and the wood. This, combined with shrinkage, tends to cause encased knots to be knocked out by rough handling or machining.

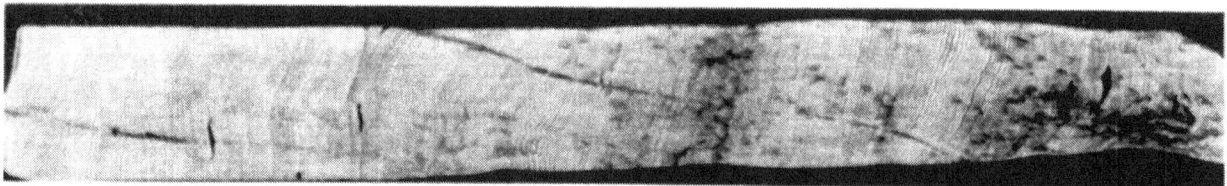

M 134 998

Figure 112.—This section from an air-dried western redcedar board shows collapse. Some ring separation is also present.

Warp

The differences in tangential and radial shrinkage characteristics of wood result in distortions of the cross section of a board. These distortions are termed warp and in lumber items are classified as cup, bow, crook, twist, and kink (fig. 113). Normal longitudinal shrinkage of wood does not greatly influence lengthwise distortions, but if juvenile and reaction wood are present, the resulting changes can cause degrade.

A general classification of the warping tendencies of some woods follows:

Tendency to warp

Low	Intermediate	High
	SOFTWOODS	
Cedars	Baldcypress	
Pine, ponderosa	Douglas-fir	
Pine, sugar	Firs, true	
Pine, white	Hemlocks	
Redwood	Larch, western	
Spruce	Pine, jack	
	Pine, lodgepole	
	Pine, red	
	Pine, southern	
	HARDWOODS	
Alder	Ash	Beech
Aspen	Basswood	Cottonwood
Birch, paper, and		
sweet	Birch, yellow	Elm,
		American
Butternut	Elm, rock	Sweetgum
Cherry	Hackberry	Sycamore
Walnut	Hickory	Tanoak
Yellow-poplar	Locust	Tupelo
	Magnolia, southern	
	Maples	
	Oaks	
	Pecan	
	Willow	

The general kinds of warp are defined as follows:

Cup—The distortion of a board in which there is a deviation flatwise from a straight line across the width of the board.

Bow—The distortion in a board that deviates from flatness lengthwise but not across its faces.

Crook—A distortion of a board in which there is a deviation edgewise from a straight line from end to end of a board.

Twist—A distortion caused by a turning or winding of the edges of a board so that the four corners of any face are no longer in the same plane.

Kink—Usually an edgewise deviation caused by a knot or severe localized grain distortion to develop two straight portions in the piece at a large obtuse angle.

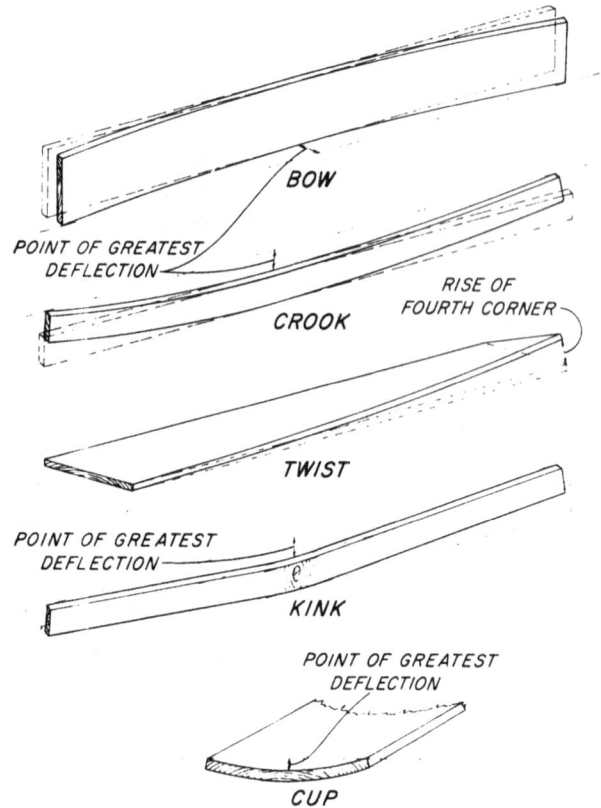

M 133 580

Figure 113.—Various types of warp in lumber.

Cup is common in plainsawed boards, and all such boards tend to cup if permitted to dry and shrink without restraint. In a plainsawed board, because of the position of the growth rings, the outer face has a greater shrinkage potential than the face nearer the pith. Consequently, when the board shrinks, the outer face tends to become concave and the opposite face tends to become convex.

For a given species, the tendency to cup varies inversely with the distance the annual rings in the board are from the pith of the tree. The other forms of warp—bow, crook, and twist—may be caused by spiral or diagonal grain, or localized distortions of grain, as around knots, for example. Spiral grain is present in the log while diagonal grain is developed by sawing crooked logs, or by sawing parallel to the pith rather than to the bark.

The amount and character of warp depend on the slope of the grain and the location and size of the areas of cross grain. Bow, crook, or twist may also be caused by the presence of bands of compression, tension, or juvenile wood that have abnormal longitudinal shrinkage. Wood near the pith, which may contain juvenile or compression wood, often has abnormal longitudinal shrinkage.

This is a good reason for boxing the heart in sawing so that the pith does not fall on one edge or face. Occasionally, the wood in the outer portions of mature trees is lighter in weight than that nearer the pith, and shrinks more along the grain; consequently, boards containing both types of wood tend to warp during drying.

Kink is observed in softwood dimension lumber such as 2 by 4's or 2 by 6's where a knot or the grain distortion that surrounds a knot is located on one edge near the middle of the piece. The two portions of the piece are essentially flat and straight. Sometimes the sharp bend in the faces of a board caused by an offset sticker and considerable superimposed loading is also called a kink.

Warp Reduction

Warping is not so sensitive to drying conditions as checking, but occurs because of the differences in shrinkage in the three grain directions, and because of irregular and distorted grain. Warping can be minimized by following good stacking and piling practices. The tiers of stickers and other supporting members, such as foundation cross beams and bolsters, should be in good alinement. Protecting the boards in the top course of a pile with a roof reduces exposure to the weather with its alternate wetting and drying cycles. Good sawing practices produce boards of uniform thickness, which is a desirable quality for stacking.

Presurfacing Green Lumber

The more refractory woods, like oak and beech, are very difficult to air dry during periods of good drying weather without surface checking. These failures result when the tension forces across the grain tend to localize at the small fractures created by the sawing process. Surfacing the green wood removes the sawing fractures, and the likelihood of surface checking in air drying is greatly reduced. By minimizing surface checking in these refractory woods, honeycombing due to penetrated and extended surface checks is literally eliminated. By presurfacing lumber to a uniform thickness, not only is the usable volume of a unit package increased, but the amount of warp present when the package or pile is razed is significantly reduced.

Clamping

Restraint to the distortion of lumber as it dries and shrinks is obtained through superimposed loading and good stickering practices. The upper unit packages in a pile provide the weighting for the lower packages. The upper courses in hand-built piles weight the lower courses. But the top package in a pile or the upper courses of lumber in hand-built piles are not loaded and distortions can take place. Clamping devices show promise in reducing the distortion problem for both air-drying and kiln-drying operations. The package hold-down or clamping device illustrated in figure 114 is quite inexpensive and effective. When two or three are used in the upper 10 to 15 courses of lumber, warping is substantially reduced. At plants where unit packages are air dried on a forklift yard and then kiln dried in either a package-loaded kiln or a track-type kiln, the hold-down devices stay with the package.

Defects Caused by Poor Stacking, Improper Piling, and Inadequate Pile Protection

During air drying of lumber the depreciation in grade and the reduction in footage due to poor stacking and piling practices and inadequate pile protection can be appreciable. Warp is aggravated. Footage losses due to end checks, end splits, and surface checking are increased.

Stacking Effects

Lumber of uniform thickness and trimmed for length can be stacked in such a way that warp is restrained. As a result of poor sorting, stacking, and piling practices, however, good grade and footage recovery might not be attained. In stack-

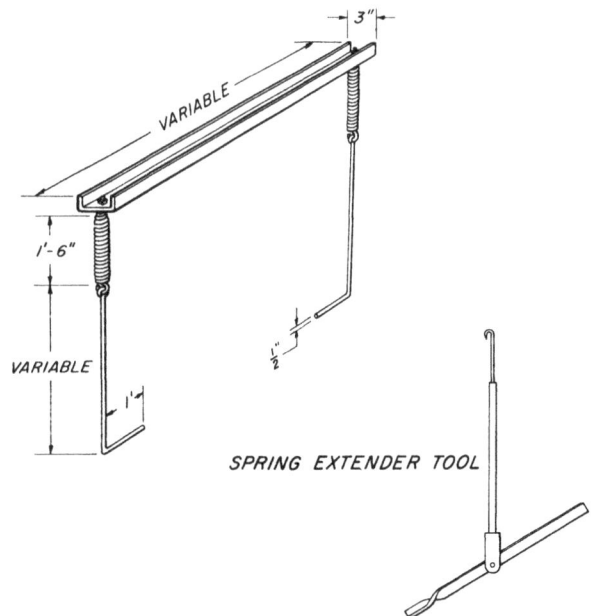

M 133 581

Figure 114.—Hold-down clamp design to restrain warp.

85

ing, the stickers, though dry, might vary considerably in thickness; when associated with poor sticker tier alinement, the introduced warp may be significantly increased. Perhaps length sorting cannot be justified, requiring that the unit packages or hand-built piles be box-piled. Insufficient sticker tiers and carelessly locating them may further fail to provide adequate board-end support. Along with careless board placement this further invites warp (fig. 115). When long boards for the edges of the box-piled courses are not handy, shorter boards are often laid down, causing sticker overhang. If the space is not blocked, sag can be created in the courses of lumber above. This is a problem of greater concern in stacking the narrower unit packages.

When stickered unit packages of lumber, particularly 1-inch softwoods of the less dense species, are transported by forklift trucks or carriers, two extra tiers of stickers are often included (fig. 48). Such tiers, located so they will be above the pickup points of the fork, are often installed for six to eight layers from the bottom. The package is usually stiffened enough so that the end stickers will not fall out during transport. Carrier bunks can be spotted at these extra sticker tiers to minimize the sag in the bottom lumber layers of the package when the package is transported or temporarily stored (fig. 116).

Figure 116.—Extra sticker tiers at the bunk locations would minimize this lumber distortion and sticker fallout.

Package Piling Effects

Unit packages are usually piled one above the others to heights limited by the elevating mechanism on the forklift truck. Thus, the lower packages are weighted by the upper ones. Any failure of the pile foundation, or the bolsters between the packages, to adequately support the load will warp the lumber. Green lumber can be bent and, if it dries while bent, the bends become fixed and are quite permanent. Where the number of cross beams in the foundation does not adequately support the pile at the sticker tiers, the lower courses are bound to sag. The introduced bow can result in appreciable grade reduction.

Figure 115.—Poor sorting and stickering practices invite warp.

Weathering Effects

When piles are not covered, the upper layers are exposed and defects develop that cause value losses due to grade reduction and footage losses in ripping and trimming to maintain grade. The economic effect is particularly notable in the upper grades of hardwoods and finish grades of softwoods. A raintight roof with overhang on sides and ends will prevent excessive moisture regain and reduce the overall drying time on the yard. Where the output of air-dried lumber per acre of yard area must be fairly high and production fairly constant, prevention of moisture regain during rainy seasons of the year is a necessity. Here good prefabricated pile covers are economical. The prevention of alternate wetting and drying also reduces grade losses caused by general weathering, mainly surface checking. The cost of the pile roof construction, placement, tiedown, storage when not in use, and repair may justify the consideration of designing and installing drying sheds, particularly for high-value lumber.

In arid areas the pile roofs should be sun tight in order to minimize grade losses.

Sun shields are often used to protect the ends and sometimes the sides of lumber piles from direct exposure to the sun. Thick material, for example, may check on the edges exposed to direct sunshine. End checking can be prevented by applying end coatings, and this is usually done rather than attaching sun shields to the ends of the packages or piles. End coatings are applied after the package is made up. When packages are ragged on both ends, as is often the case with redwood, the application of end coatings is difficult. Sun shields are used at some redwood sawmills rather than attempt end coating.

The movement of air in very open yards may need to be slowed during hot, dry weather to prevent excessive surface checking in some woods. The windward sides of the piles can be covered with tarpaulins or sheets of polyethylene to act as wind barriers.

For Additional Information

Cech, M. Y.
    1966. New treatment to prevent brown stain in white pine. Forest Prod. J. 16(1): 23–27.

Clark, J. W.
    1957. Comparative decay resistance of some common pines, hemlock, spruce, and true fir. Forest Sci. 3: 314–320.
Cuppett, D. G.
    1965. How to determine seasoning degrade losses in sawmill lumber yards. USDA, Forest Serv., Northeast. Forest Exp. Sta. Res. Note NE 32, Upper Darby, Pa.
Hunt, G. M., and Garratt, G. A.
    1967. Wood Preservation. Ed. 3. McGraw-Hill, N.Y.
McMillen, J. M.
    1961. Coatings for the prevention of end checks in logs and lumber. USDA, Forest Serv., Forest Prod. Lab. Rep. 1435, Madison, Wis.

———
    1963. Stresses in wood during drying. USDA, Forest Serv., Forest Prod. Lab. Rep. 1652, Madison, Wis.
Page, R. H., and Carter, R. M.
    1957. Heavy losses in air seasoning Georgia pine and how to reduce them. USDA, Forest Serv., Southeast. Forest Exp. Sta. Pap. 85, Asheville, N.C.
Peck, E. C.
    1954. Effects of machine stacking on drying rate and degrade. South. Lumberman 189(2362): 62–67, Sept. 1.

———
    1957. Air drying and sticker staining of 4/4 sugar maple flooring stock in upper Michigan. USDA, Forest Serv., Forest Prod. Lab. Report 2086, Madison, Wis.
Scheffer, T. C.
    1958. Control of decay and sap stain in logs and green lumber. USDA, Forest Serv., Forest Prod. Lab. Rep. 2107, Madison, Wis.
Stutz, R. E.
    1959. Control of brown stain in sugar pine with sodium azide. Forest Prod. J. 9(12): 459–463.
———, Koch, P., and Oldham, M. L.
    1961. Control of brown stain in eastern white pine. Forest Prod. J. 11(6): 258–260.
U.S. Forest Products Laboratory
    1958. Cause and prevention of blue stain in wood. USDA, Forest Serv., Forest Prod. Lab. Tech. Note 225, Madison, Wis.

———
    1951. Chemical brown stain in pine. USDA, Forest Serv., Forest Prod. Lab. Tech. Note 254, Madison, Wis.

CHAPTER 7

FACTORS AFFECTING THE COST OF AIR DRYING LUMBER

Several studies have been made to determine the cost of air drying lumber. While the cost data developed are good for the yards studied, they are of limited value in trying to arrive at a meaningful general cost figure. The factors involved are too numerous and complex, and there are no standard criteria for evaluating these costs. For example, some operators charge lumber sorting to the operation of the sawmill; others count it a part of the cost of air drying. Some add degrade to air-drying costs; others do not. Some plant owners charge the cost of training lumber inspectors to general overhead; others to lumber drying. Insurance, taxes, and interest charges on inventory are usually but not always considered a cost of the air-drying function. These are only a few of the differences in cost accounting at the many drying operations throughout the country.

Actual air-drying costs also vary with many factors. These include lumber species and thickness, weather, yard site and arrangement, lumber handling and piling methods, lumber volume, initial and target moisture content values, yard output per acre, and losses of grade and footage caused by drying defects. Other factors are the costs involved in the purchase or lease of land and its improvements, and the amount invested in pile foundations, roofs, stickers, sun shields, road maintenance, drainage, equipment purchase, upkeep, and depreciation. Labor costs and taxes, insurance, and interest on yard inventory often vary considerably from plant to plant.

Yarding costs, or that part of the cost of air drying lumber that can be attributed to the operation and maintenance of the yard, have been estimated as low as 5 cents per day per 1,000 board feet for softwoods to as high as 25 cents for hardwoods.

Data from several studies are presented as examples of accounting procedures used to arrive at the cost of air drying lumber. Listed at the end of this chapter are the published references that were used as a basis for some of the cost information presented here. Other cost data have been gathered from unpublished sources.

Example 1.—A study by D. G. Cuppett of the Northeastern Forest Experiment Station involved 10 eastern hardwood sawmills. Mill production varied from a small operation cutting 500,000

board feet (500 MBF) per year to one producing 6 million board feet per year. Tram or dock yards with hand-built piles and forklift yards are both included in the following summary:

Average cost to air dry hardwood lumber at these 10 sawmills:

Item	Average cost per MBF (dollars)	Percent of total
Variable costs:		
Labor and payroll taxes	8.13	57.5
Maintenance and repairs	.98	6.9
Supplies	.44	3.1
Miscellaneous	.66	4.7
Fixed costs:		
Depreciation	2.16	15.3
Interest	1.41	10.0
Real estate and property taxes	.25	1.8
Other taxes (except income)	.10	.7
Total	14.13	100.0

These costs do not include air-drying grade losses, which may range from 2 to 10 percent or more of the lumber value. While insurance is not listed, it is probably included under miscellaneous costs.

Example 2.—Hardwood air-drying cost data at a midwestern custom kiln-drying operation were reported some years ago by E. M. Conway of Conway Corp., Grand Rapids, Mich. Again, grade loss was not included in this report. Interest on the capital investment was not included and no inventory costs were considered because the lumber was not owned by the operator. The following summary gives the expense of lumber handling, yarding, and unstacking:

Item	Average cost per MBF (dollars)	Percent of total
Labor (plus supervision, fringes, and payroll taxes)	6.20	64.3
Depreciation and amortization of handling equipment	1.49	15.4
Maintenance of handling equipment	.95	9.9
Insurance	.34	3.5
Land and buildings expense	.31	3.2
Other (gasoline, demurrage, cleanup, cartage, etc.)	.36	3.7
Total	9.65	100.0

In each of these examples, labor and depreciation costs in the yarding operation were well over 70 percent of the total cost. The cost per day based on an average 90-day air-drying period would be 15.7 cents per day per 1,000 board feet for the eastern sawmills and 10.7 cents per day per 1,000 board feet at the midwest plant.

Example 3.—In another hardwood air-drying study [3] the following costs of operation were summarized:

Item	Cost per MBF (dollars)	Percent of total
Inspection	1. 40	9. 4
Yard labor	6. 15	41. 2
Maintenance	. 60	4. 2
Supplies	. 50	3. 4
Fuel, oil, grease	1. 05	7. 0
Payroll taxes	. 75	5. 1
Depreciation	2. 00	13. 3
Interest	1. 45	9. 7
Insurance	1. 00	6. 7
Total	14. 90	100. 0

If an average 90-day drying period is assumed, the cost of air drying is 16.5 cents per day per 1,000 board feet. Drying degrade, however is not included in this analysis.

Example 4.—At a redwood operation using a pile spacing of 6 feet and a drying time of 12 months, the air-drying cost was $13.24 per 1,000 board feet.[4] The costs of sorting and unstacking were not charged against the yarding operation. Also, grade and footage losses were not included in the following tabulation:

Item	Average cost per MBF (dollars)	Percent of total
Stickering and transport	1. 82	13. 8
Lumber inventory cost	9. 64	72. 8
Yard costs	1. 78	13. 4
Total	13. 24	100. 0

The air-drying cost in this example is 3.6 cents per day per 1,000 board feet.

Relationship of Degrade and Total Cost

A major item not always included in air-drying cost analysis is degrade. It oftens amounts to a large part of the overall cost of drying lumber; by minimizing degrade, a substantial savings can usually be realized.

Studies have been made of drying grade losses in both softwoods and hardwoods. Less attention has been given volume loss associated with drying. However, this factor must be considered, where applicable, in computing drying costs. The results

[3] This information was part of a lecture given at a kiln drying course at the Forest Products Laboratory in April 1966 by D. G. Cuppett of the Northeastern Forest Experiment Station.

[4] This information was presented at a kiln drying short course in 1964 by Fred Dickinson of the University of California Forest Products Laboratory, Richmond, Calif.

of two degrade studies are described here to show the extent of air-drying losses.

Example 5.—The effect of different piling methods on grade loss in 4/4 southern pine lumber was studied by Roy Carter of North Carolina State University and Rufus Page of the Southeastern Forest Experiment Station. The average loss for all piles was $10.21 per 1,000 board feet. Losses ranged from $5.21 for forklift yarding to $26.21 for hand-built piles. Stain accounted for the major part of these losses, followed by crook, bow, decay, and surface checks, with end checks and cup last. Based on this study, a yard drying 2 million feet of 4/4 pine in the South might expect to lose over $10,000 annually; one drying 5 million feet, more than $26,000; and one drying 10 million feet, in excess of $50,000 per year from degrade incurred in air drying alone.

Example 6.—Cuppett's study (see example 1) of air-drying grade losses per 1,000 board feet of 4/4 red oak lumber in the Central Appalachians indicated losses of $25.73 on small yards, $9.90 on large yards, and an average loss of $17.78 for all yards sampled. Surface checks accounted for more than half the loss, followed by splits, warp, and stain and other defects. Both surface checks and splits were more prevalent in stacks that were not covered. Most warp occurred in piles where stickers and bolsters were poorly alined. Average grade loss for lumber piled in the fall and winter was $18.78, compared with $5.18 per 1,000 board feet for lumber piled in the spring and summer.

Using the cost figure developed at the 10 eastern hardwood mills, $14.13, and adding the grade loss occurring on large yards cited above, $9.90, the total hypothetical cost of air drying eastern hardwood lumber is $24.03 per 1,000 board feet, of which 41 percent is from degrade. If we use the figure for average grade loss, $17.78, the total cost would be $31.91, with 55 percent of this attributable to drying degrade. This would be an actual cost figure for the 10 hardwood mills if grade loss had been evaluated at these same mills.

An Air-Drying Cost Equation

An ingenious formula for determining air-drying costs was devised by D. D. Johnston of the British Forest Products Research Laboratory. Component costs are combined in a cost equation to arrive at total drying costs, exclusive of handling costs and degrade loss, as follows:

$$\frac{t}{K}[(C+L+T)r+C(x+y)+T(z)]$$

where $t=$ mean drying time in years
$K=$ yard capacity
$C=$ capital cost of establishing a drying yard (including site preparation, roadways, foundations, stickers, and pile covers)
$L=$ land value
$T=$ value of lumber being dried
$r=$ current interest rate (value/100)
$x=$ depreciation (value/100)
$y=$ maintenance (value/100)
$z=$ insurance and office expense (value/100)

Taking a hypothetical case, using dollars and board feet where the author used pounds and cubic feet, and applying the cost factors with the following values:

$t=9$ months	$T=\$300,000$
$K=2$ million board feet	$r=0.06$
	$x=0.15$
$C=\$30,000$	$y=0.10$
$L=\$5,000$	$z=0.01$

The equation becomes:

$$\frac{0.75}{2,000,000}[(30,000+5,000+300,000)0.06$$
$$+30,000(0.15+0.10)+300,000(0.01)]$$
$$=0.000000375(20,100+7,500+3,000)$$
$$=0.000000375\times30,600=0.011475\text{¢ per board}$$

foot or $11.48 per 1,000 board feet.

If we assume a handling cost of $7 per 1,000 board feet, this brings the total drying cost, exclusive of degrade loss, to $18.48. Tacking on another $11 per 1,000 board feet for grade loss, the total air-drying costs, in this hypothetical case, would be $29.48 per 1,000 board feet.

The above formula could be tested by interested operators using actual component costs. Results could then be checked against those obtained by conventional accounting methods and modified to fit the situation. If it proves satisfactory, it would be a handy tool for calculating air-drying costs for any given period of time.

Controllable Cost Factors

A number of cost factors in air drying can be controlled to some extent. The most important are drying time, drying degrade, and labor.

Drying Time

A major component of air-drying costs is interest paid on inventory value. Many plant owners who operate on borrowed money are acutely aware of this. Some operators report inventory costs as high as 75 percent of total costs. Thus, a reduction in total costs is dependent, to a large degree, on reducing the time required to dry the lumber. As an example, if a plant air dries 5 million board feet of 4/4 oak annually at a cost of 15 cents per thousand per day, a total annual savings of $22,500 would be realized by reducing yard time by 30 days.

Because drying costs are so closely related to drying time, the dense hardwoods, which are often more valuable and which dry more slowly than light hardwoods and most softwoods, cost more to dry. Drying time can be optimized by controlling the factors that affect the rate of drying, such as pile orientation, yard layout, yard surface, yard sanitation, pile dimensions and spacing, pile roofs and foundations, lumber size, and sticker thickness.

Degrade

The previous examples of grade loss in both pine and hardwoods suggest the importance of controlling the factors that affect the degree and kind of degrade. Many of these factors are also involved in rapid drying.

Proper pile construction, which includes a good foundation, well-alined stickers and bolsters, and a satisfactory roof, is a major factor in controlling degrade. Foundations that are too low or prevent escape of moisture-laden air from beneath the piles contribute to the high cost of air drying lumber, for yard time is increased.

A very important item in minimizing costly degrade is the use of rain- and sun-tight pile roofs for lumber dried in the open. Heavy losses in air drying often occur in the top two or three courses of uncovered piles because they are unrestrained and unprotected. In the study by Page and Carter (Example 5) of mixed grades of 4/4 southern pine lumber, losses from drying in a forklift yard, by pile section, were as follows:

	Cost per MBF
Top section (top 3 courses)	$11.51
Middle section	4.75
Bottom section (bottom 2 courses)	9.31

Example 7.—A study by Carter and Page to evaluate pile roofs was made in the South. This involved 4/4 No. 1 Common yellow-poplar and black and tupelo gum. Applying the data to commercial operations, they showed the value loss in lumber yield to be $21,000 in the top four layers of yellow-poplar for a plant drying 5 million board feet per year, and for gum, a more refractory

species, about $28,000 for a yard handling 3.75 million board feet annually.

Example 8.—At an oak flooring plant,[5] the owner calculated the savings from using pile roofs as follows:

	Savings per MBF
In volume: 80 board feet saving in yield per pile at $100 per MBF	$8.00
In grade: 100 board feet saving in yield per pile from degrading 1C to 3A ($125 to $75 per MBF)	5.00
	$13.00

At this plant, pile roofs save $22,000 annually, less roof costs of approximately $1,300 per year. This leaves a substantial savings from covering the piles.

Stickers

A part of the cost of yard operation is in stickers. This cost factor was considered in a report by L. D. Brown of Brooks-Scanlon, Inc., Bend, Oreg. Sticker costs at a large western mill were 22 cents each for those made from vertical-grain, D and Better Douglas-fir, and 27 cents for those made from pine pitchy Selects. At this plant, 200 stickers per day were taken out of service. Ninety-nine percent of these were repaired at a cost of 11.7 cents each. This included the cost of cutting and gluing and supplies such as glue and staples. Many plants have found sticker repair profitable.

Another sticker study involving 12 western mills was reported by the Washington-Idaho-Montana Kiln Club. The average cost per long sticker (96 to 108 inches) was 19.7 cents and per short sticker (40 to 54 inches) 6.4 cents. This same study concluded that Select grades of lumber produce more and better stickers than Common grades. Their use was justified despite greater material costs. Stickers an inch or more in thickness gave longer service life and cost less in the long run than thinner stickers.

Labor

Labor costs are controllable to the extent that most operations can be more completely mechanized. However, it is important to recognize that overinvestment in equipment designed to replace labor has been the downfall of many operators. Each operator must determine the point at which mechanization is no longer feasible. Machine operators such as head sawyers and edgermen, as well as other skilled labor, should be retrained periodically. Training pays off in better lumber and in increased production rates. Providing adequate training for even unskilled labor has proved profitable at many plants.

Insurance

The cost of workmen's compensation insurance is influenced substantially by the firm's safety record, and is usually reevaluated periodically. The safer the operation, the less this insurance costs. Insurance of lumber piled on the yard is generally less for a clean yard with adequate facilities for fighting fire and with alleys and roadways wide enough to serve as firebreaks.

Maintenance

Care in operating equipment and close attention to preventive maintenance will, in most cases, substantially reduce equipment operating costs. Yard maintenance costs will be less if the yard is well drained and paved. Also, a well-surfaced yard is safer and it is less expensive to operate lifts and straddle trucks.

For Additional Information

Anonymous
    1963. Automated lumber yard cuts costs. Furniture Design and Manufacture 35(4):25.

Brown, L. D.
    1964. Dry kiln sticker repair costs and procedures. Forest Prod. Res. Soc., Wood Drying Div. News-Digest, Apr.

Carter, R. M.
    1957. Uncover seasoning losses and cover seasoning yards. Forest Prod. Res. Soc., Wood Drying Div. News-Digest, Feb.

Conway, E. M.
    1958. Factors that affect the cost of hardwood kiln drying. Forest Prod. J. 8(1):21A–24A.

Cuppett, D. G.
    1966. Air-drying practices in the Central Appalachians. U.S. Forest Serv. Res. Pap. NE 56. Northeast. Forest Exp. St., Upper Darby, Pa.

Johnston, D. D.
    1967. A note on the cost of air drying. Timber Trades J. 260(4715):53–54.

Minter, J. L.
    1965. Savings prove value of mechanized yard. Wood and Wood Prod. 70(7):27.

Washington-Idaho-Montana Kiln Club
    1957. Sticker cost study. Forest Prod. Res. Soc. Wood Drying Div. News-Digest, Sept.

[5] From data collected by R. Page at the Southeastern Forest Experiment Station in 1957.

CHAPTER 8

PROTECTION OF AIR-DRIED LUMBER

When lumber is as dry as can be expected on the air-drying yard, it should be stored where it is no longer exposed to wind, sun, and rain. The deteriorating effect of weathering continues as long as lumber is left outdoors. When stickered lumber is taken down and solid stacked, it takes up less space and a shed can provide the needed protection. When yard space is needed for air drying lumber, less favorable drying areas are often set aside for bulk storage of lumber that has already been air dried.

Naturally, the seasonal changes that influence the air-drying potential of an area also affect the conditions of dry lumber storage. But significant moisture changes in air-dried lumber result mainly from rains that rewet the lumber. If good, rain-tight pile roofs with adequate projections were used during drying or if the lumber were shed dried, the needed protection from adverse weather was provided.

Lumber that has reached 20 percent moisture content or less is usually no longer threatened by the development of stains or fungus and insect attack. Consideration may be given to storing and protecting the lumber without further drying.

Outdoor Storage

The air-drying process is, in effect, outdoor lumber storage. The inventory is handled in such a way that drying takes place during the storage period. But the inventory period on the air-drying yard should be no longer than that required to reduce the moisture content to a level suitable for shipment. Then it becomes a situation of protecting the air-dried lumber against rewetting until shipping orders are received for the particular species, size, and grade. This protection is especially important if the lumber is of high grade and the storage period is likely to be long. Roofs should now be provided if they were not when the pile was first erected. The lumber may be left on sticks, or the piles and packages may be razed and the lumber bulk piled for more efficient use of space.

If the average moisture content of the lumber is greater than 20 percent, it should be protected in such a way that drying can continue, as under a roof or in a shed; it should not be bulk piled because of the hazards of stain and decay development. When the moisture content of the lumber is less than 20 percent, the piles or packages may

be wrapped with plastic tarpaulins or enclosed in prefabricated waterproof-paper shrouds (fig. 117) for temporary outdoor storage.

Foundations in the storage area should provide good support and ground clearance just as in the original air-drying pile. Support is still essential to prevent pile sagging. Ground clearance is not only needed to provide room for forklift operation but also for ventilation to keep moisture regain in the bottom layers to a minimum. If the water table in the ground is high, a ground cover to provide a barrier to moisture vapor movement out of the soil may be desirable, particularly if the storage period is extended. Precautions should be taken to protect the lumber against infestation by termites and other insects.

Indoor Storage

The indoor storage of air-dried lumber may be provided by buildings called sheds and more specifically "dry sheds." The advantage of shed storage is the protection provided by the permanent roof. The air-dried lumber is stored in the shed either as stickered unit packages, as bulk-piled lumber in unit packages, or bulk piled in bins or bays. The sheds can be open, closed, or closed and heated. They may or may not be floored or paved. Their design, size, and materials of construction will vary depending upon plant arrangement, storage-volume requirements, and the cost of locally available building materials.

Open-Shed Lumber Storage

An "open" lumber storage shed is one that does not usually have sides or ends to block air movement through the building (fig. 118). Sheds with louvered walls are considered open sheds (fig. 119). Wind-blown rain is baffled, a distinct feature of sheds of this type. If the shed is not floored or paved, foundations for the stickered or bulked packages of air-dried lumber will be necessary to provide ground clearance for ventilation.

When hand-built piles in the yard are taken down, the off-grade lumber is often bulk piled in bins or bays in the "rough dry shed." In a forklift-yard operation, the stickered unit packages or packages of bulked lumber are stored in the open shed; package pile height is governed by the height

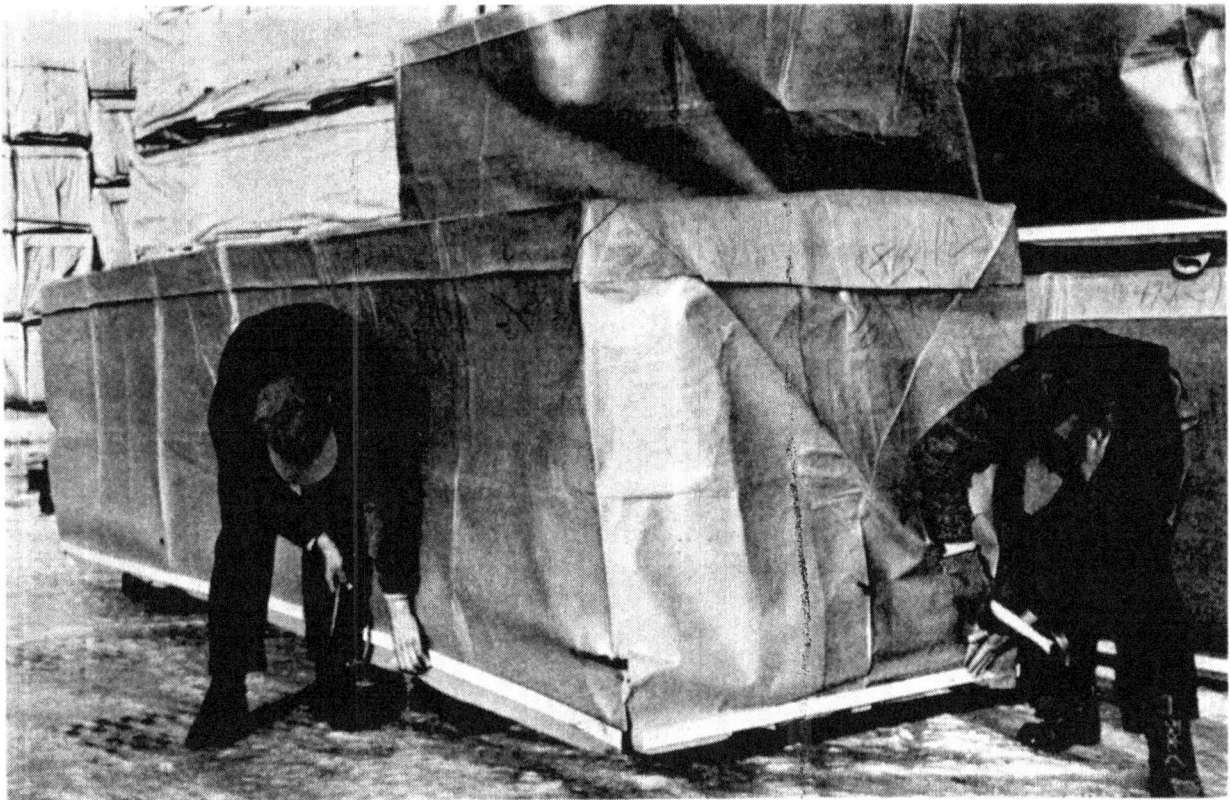

Figure 117.—A prefabricated waterproof-paper shroud is being attached to a package of bulk-piled dry lumber to protect it during temporary outdoor storage.

of the shed (fig. 120). Lumber storage volume can be increased by having a forklift available to place carrier packages two, three, or more units high, depending upon the height and construction of the shed.

Closed, Unheated Lumber Storage Shed

A closed, unheated shed for the storage of air-dried lumber differs from an open shed in that the sides and ends of the building are walled,

Figure 118.—Open shed for storage of dry lumber.

93

Figure 119.—Open sheds with louvered side walls are frequently used for the storage of air-dried lumber.

blocking air movement through the structure (fig. 121). Doorways and other access openings for the transport of lumber in and out of the shed are usually kept closed when activity is not great or the sawmill or factory is not in operation. Ventilators are sometimes installed on the roofs of these buildings but the air circulation within the shed remains virtually stagnant. Partially dry, stickered units of lumber can be stored in a closed shed but, if further drying at a reasonable rate is wanted, the air

94

Figure 120.—Open shed for storage of dry lumber. Both bulk-piled and wrapped unit packages are stored in the shed.

movement through the stickered package must be stimulated by introducing forced-air circulation with fans.

As air movement in and out of a closed shed is restricted by the solid walls, some heating of the air in the building results from solar radiation.

If the roof and walls exposed to sunshine are dull black, these structural elements are heated when the sun shines and the air inside the building in contact with these heated elements is warmed. The air in the upper space of the building is warmer than the air at the ground level and, if the warmed

Figure 121.—A closed, unheated lumber storage shed.

air is moved downward by fans, the EMC conditions of lumber storage are lowered. Air-dried lumber is more certain to remain dry, and partially air-dried lumber in stickered unit packages will dry faster.

If the closed shed is not floored or paved, foundations must be provided for the unit packages or for the lumber in the bins in order to create a space between the ground and the lumber. The walls of the building are often open near the ground level in a closed shed of this kind to create air drift or ventilation under the lumber. These ground-level wall openings can be screened or louvered to discourage entry of birds and small animals into the shed.

Closed, Heated Lumber Storage Shed

A closed, heated shed is most often utilized to store lumber that was kiln dried to moisture content levels lower than can usually be attained in conventional air drying. However, such a shed might sometimes be used to dry stickered unit packages of air-dried lumber even more. After all, the systems used to heat the closed shed involve forced-air circulation with fans.

The closed, heated storage shed is usually floored or paved. The unit packages of lumber, or bulked lumber in bins, should not be in direct contact with a paved floor, however, as some moisture might migrate through the paving. The warmed air being circulated in the shed should be allowed to move under the stored lumber unit; in most instances the bunks used in carrier or forklift operations provide adequate elevation of the lumber. The temperatures maintained in a closed, heated lumber storage shed are not very high, and insulated walls, roofs, or ceilings are seldom installed.

The wood EMC conditions in a heated storage shed depend upon the outdoor temperature and relative humidity and the temperature of the circulated air within the shed. When outdoor air is heated, the absolute humidity remains the same but the relative humidity is lowered. Thus, the wood EMC of outdoor air is reduced. In figure 122, the wood EMC lines of the chart are almost parallel with the horizontal relative humidity lines. The chart is used to determine the temperature that should be maintained in the heated storage shed to prevent dry wood from absorbing moisture.

For example, if during the winter months the outdoor temperature averages 40° F. and 75 percent relative humidity, the wood EMC is 15 percent. If the shed is to be heated so that the wood EMC is not less than 12 percent, what temperature

should be maintained? The absolute humidity at 40° F. and 75 percent relative humidity is about 2.2 grains per cubic foot. Follow the 2.2-grain imaginary line (parallel to the 2-grain line) down and to the right until it intersects the 12 percent EMC line. Then read the temperature at the bottom of the chart at this intersection point. It is about 45° F. In this case, by heating the air to 45° F., the wood EMC is lowered from 15 percent to 12 percent.

If kiln-dried lumber at 6 percent was stored in the shed, the temperature would have to be increased to about 65° F. to maintain the 6 percent EMC condition under the above winter weather conditions.

Air-dried lumber in storage does not require much heating of the ambient air to prevent moisture regain, and it is quite possible that in many areas of the country solar radiation will provide the needed heat energy. Circulation of the air within a closed, heated lumber storage shed by means of fans is necessary regardless of how heating is accomplished, particularly in a solar-heated system. The fans move the air across the warmed roof and walls and mix this warmed air with that being circulated around the dry, stored lumber.

Closed sheds at millwork and furniture plants are often equipped with unit heaters complete with fans. Most often these unit heaters are supplied with steam from a central plant, but individual gas-burning units are sometimes used. The heaters are located in such a way that the warmed air is directed toward the stored lumber. The conventional temperature control of a unit heater is a thermostatic control of the fan motor. When the thermostat calls for heat, the unit heater fan motor is started and air is forced through the finned coils of the heater. As forced-air circulation in the shed is desirable at all times, the unit heater fan motor can be kept in full-time operation and the temperature controlled by installing a motor valve on the steam or gas supply line for "on and off" operation. This control mechanism is more expensive to install than a thermostatically controlled fan but it does insure full-time fan operation to keep the warmed air in the heated shed from stagnating and stratifying.

As temperature control in a closed, heated shed for dry lumber is mainly a problem of controlling wood EMC conditions at a predetermined value, the most simple control system is a humidistat which controls the shed temperature. By controlling the relative humidity in the heated shed, the wood EMC condition is controlled closely enough for most practical purposes. For example, from figure 122, the relative humidity that establishes

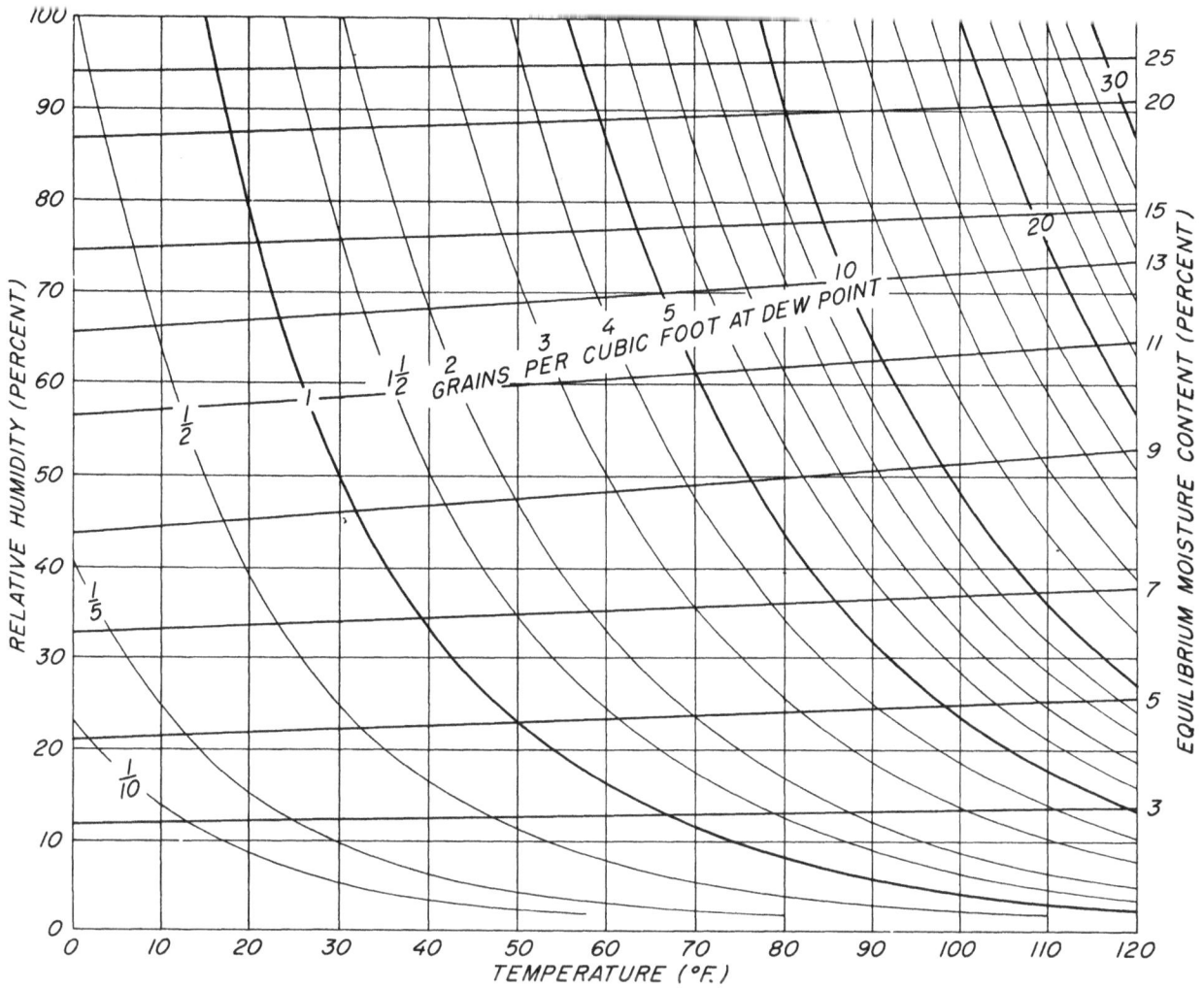

M 133 676

Figure 122.—Equilibrium moisture content of wood as a function of temperature, relative humidity, and absolute humidity.

a 12 percent wood EMC condition is about 65 percent over a fairly wide range of temperature. By setting the humidistat at 65 percent relative humidity, the indoor air will be heated to lower the indoor relative humidity to this value. The advantage of a humidistat control installation on the unit heaters is that, with changing outdoor temperatures and relative humidities, the wood EMC in the heated shed remains substantially constant. With humidistat control, the fans on the unit heaters should operate continuously and the humidistat should control the steam or gas input into the heater.

The hazards of stain and decay developing in the lumber stored in a heated shed are significantly reduced. Partially air-dried lumber, however, should be left on sticks so that drying can continue in storage.

Protection of Air-Dried Lumber in Transit

Air-dried lumber is shipped from the producing sawmills by railroad and truck to the factories where it may be kiln dried to a lower moisture content for final processing. The air-dried rough lumber, however, may be kiln dried and processed to some extent at custom kiln-drying companies who carry out these services on an "in transit" basis. When air-dried lumber is shipped by the producer, either by railroad or by truck, protection from the weather may or may not be provided.

97

Railroad Transit

High-value, air-dried lumber is often shipped to the wood-using factories in boxcars. The lumber is thus protected from the weather, and deterioration of any kind is not likely to occur. The labor costs of loading and unloading have affected the volume of shipments transported this way. To facilitate loading and unloading, boxcars are available with wide doors so that unitized packages of bulk-piled lumber can be moved in and out of the boxcar with lift trucks.

Air-dried lumber is also shipped in gondola cars. The bulk-piled lumber may be loaded into the gondola cars by hand, or unit packages can be loaded into the car by crane or by lift truck equipped with special slings. The unit packages of bulk-piled lumber are individually strapped and the completed gondola load is strapped or chained in such a manner that the lumber cannot shift during rail transit. Protection from the weather is seldom provided.

Flatcars are being more widely used for the transport of air-dried lumber as they can be quickly loaded and unloaded by lift truck. The air-dried lumber is unitized into forklift packages that are loaded onto the flatcar and anchored by strapping or other binding methods. Some railroads use "chain" flatcars for lumber shipments (fig. 123). These flatcars are loaded with unit packages of bulked lumber by lift truck. When the carload is in place, it is anchored or bound to the car by numerous chains that are thrown over the load and tightened with ratchet load binders. Metal corner protectors may be used to prevent chain damage to the edges of the boards in the top layer of the top packages in the load. These "chain" cars may or may not have bulkheads at the flatcar ends. Flatcars with "A" frames in the longitudinal center are also available and the advantage is increased sidewise stability of the flatcar load.

Truck Transport

Considerable quantities of air-dried lumber are shipped by truck from the producing sawmills to the using factories or custom kiln-drying plants. Tractor-trailer units are usually used for this purpose, and in most instances the trailer is a flatbed unit that can be loaded and unloaded by lift truck. The load of lumber is anchored to the trailer by chains tightened with load binders. High-value, air-dried lumber is often protected by covering the load with canvas tarpaulins (fig. 124). Lower grade lumber with short hauls is seldom protected.

M 135 004

Figure 123.—Chain car loaded with unit packages of air-dried lumber. Corner irons protect the top courses of lumber.

M 135 080

Figure 124.—Tarpaulin protection on a trailer load of air-dried lumber.

For Additional Information

Anonymous
   1964. Wrap it to protect it from the weather. Hitchcock's Woodworking Dig. 66(9): 92.

Applefield, Milton
   1964. Polyethylene and kraft paper wrapping for protecting kiln-dried lumber stored in the open. Hardwood Res. Counc., Statesville, N.C.

Canadian Department of Resources and Development
   1952. Moisture content changes in seasoned lumber in storage and in transit. Forest. Branch Bull. 102, Ottawa.

Gill, T. G., Robertson, D. V., and Miller, R. J.
   1956. Handbook for the preservative treatment of packaged lumber and fabricated wood products for long-term storage. Timber Eng. Co., Washington D.C.

Hickman, L.
   1958. Protection of lumber for exterior storage. Western Pine Assoc. Res. Note 6.711, Portland, Oreg.

Patterson, D.
   1958. How to design pole-type buildings. Ed. 2, Amer. Wood Preservers Inst., Chicago, Ill.

Peck, E. C.
   1955. Storage and handling of lumber. USDA, Forest Serv., Forest Prod. Lab. Rep. 1919, Madison, Wis.

Rasmussen, E. F.
   1959. Some effects of storage on seasoned lumber. USDA, Forest Serv., Forest Prod. Lab. Rep. 1071, Madison, Wis.

99

SUMMARIZED GUIDE FOR AIR-DRYING PRACTICES

1. The Air-Drying Yard

a. *Site.*—High, well-drained ground with no obstructions to prevailing winds provides the greatest drying potential. Transportation is facilitated if the yard is located close to the sawmill or woodworking factory.

b. *Yard layout.*—Roadways of main and cross alleys divide the yard into rectangular blocks or areas for the lumber piles. The row-type forklift yard has main alleys at least 24 feet wide with cross alleys that may be 60 feet wide. In a line-type forklift yard, a 60-foot cross alley might be the principal roadway from the sawmill or woodworking factory. The lines of piles and the main alleys are placed at right angles to the cross alley. The length of the lines is determined by yard boundaries or the need for cross alleys for insurance or fire fighting purposes. The hand-built pile yard is usually laid out in blocks of main alleys with a rear alley between each two rows of piles. The main alleys can be relatively narrow but the blocks are often separated by wide cross alleys that might also be fire lanes.

c. *Yard orientation.*—Main alleys often run north and south to obtain faster roadway drying after rains or faster snow and ice melt. In forklift yards the main alleys are sometimes laid out parallel to the prevailing wind to improve air movement in the unit packages. In hand-built pile yards the orientation of the main alleys with respect to the prevailing wind is not very important, if the piles are on high foundations.

d. *Row spacing.*—In a row-type yard the row space needs to be wide enough to enable the forklift operator to maneuver and place the packages on the pile. A spacing between rows of 3 feet is recommended. A closer spacing might be practical in a paved yard. A "side-shifter" on the forklift truck will make unit package placement much easier for the forklift driver.

e. *Line spacing.*—The space between the two lines in a line-type yard varies but 2 feet is considered as a minimum. The two lines of permanent foundations need to be separated by this much or more to assure unobstructed air drift between the lines.

f. *Pile width.*—Narrow packages of lumber tend to tip, often requiring long tie bolsters between piles. Wider unit packages minimize the possibility of tipping and some hardwood operations are piling 8-foot packages successfully. Hand-built piles of hardwoods are usually 8 feet wide. Self-stickered softwood piles vary from 8 to 16 feet wide, depending upon the sorted length of the lumber being stacked.

g. *Pile heights.*—Although the pile height in a forklift yard is limited by the elevating capacity of the forklift truck, a standard practice is to pile four unit packages on a foundation, making a pile height of about 20 feet. Higher piles of unit packages are not recommended. The hand-built pile, if built up from the ground, is usually about 15 feet high but those erected alongside an elevated dock may be as much as 30 feet high. Compression of stickers and the indentation of the lumber in the lower courses because of superimposed loading may require use of wider stickers in building these high piles.

h. *Pile spacing.*—The recommended minimum pile spacing in a row-type forklift yard is 2 feet. The pile spacing in the row can be varied to open up the yard when necessary. The space between ends of piles in a line-type forklift yard can be 1 foot or less depending upon whether the forklift truck is equipped with a side shifter. The recommended space between piles in a hand-built yard is 3 feet.

i. *Pile placement.*—In a row-type forklift yard the rows are placed perpendicular to the main alley. The length of the row is determined by the inventory. In a line-type forklift yard the lines are parallel to the main alleys. Hand-built piles are placed perpendicular to the main alley.

j. *Yard transportation methods.*—For easy and rapid handling, motorized transport equipment is used to move lumber from the sorter to the yard or from the stacker to the yard. For relatively long hauls carriers are more efficient than forklift trucks.

k. *Pile foundations.*—Wood foundation members are generally used for row-type forklift yards and hand-built pile yards. Preservative treatment of mudsills, posts, and piers is justified. When concrete piers are used to support stringers or cross beams, their footings should be below frostline. Concrete piers supporting railroad rails are frequently used in line-type forklift yards. The minimum foundation height for forklift yards is 8 inches and the preferred design allows free air movement in any direction under the pile. The recommended height at the low end of the sloped foundation for hand-built piles is 18 inches.

l. *Pile protection.*—Sun- and rain-tight pile roofs protect high-value lumber from weathering. Sun shields or end coatings reduce footage losses caused by end checking and end splitting in thicker lumber items. Wind barrier tarpaulins can minimize surface checking when they are used to cover piles on the windward side of a forklift yard during periods of hot, dry winds.

2. *Stacking and Piling Lumber for Air Drying*

a. *Sorting.*—Drying time varies with thickness, and yard output is greater if thickness segregations are made for stacking and selected yard placement. Sorting for length is also recommended. Stacking of lumber sorted for length reduces grade losses caused by warp. Sorting for species is generally quite essential.

b. *Sorting equipment.*—Smaller sawmills and woodworking plants will set up a conventional green chain or other mechanical or semimechanical means to sort lumber for stacking. At larger sawmills mechanical sorting equipment such as edge, drop, or tray sorters can reduce lumber-handling costs.

c. *Stacking of lumber into unit packages.*—Sticker guides on stacking stalls or mechanical stackers assure good sticker alinement. Mechanical stackers provided with an "even-ender" produce square ends on the packages. When random-length lumber is stacked, box piling is recommended.

d. *Board spacing.*—Edge-to-edge stacking of both softwoods and hardwoods is practical for unit packages 4 feet or less in width. For wider packages of random-width hardwoods, board spacing may be necessary to minimize blue stain. Flues or chimneys are built into wide unit packages of stain-prone softwoods.

e. *Package height.*—The load-carrying capacity of the lift truck often determines the unit package height. A package height of about 4 feet is recommended as a general standard for most yarding operations.

f. *Bolster size.*—Small wood timbers 4 by 4 inches in cross section and as long as the unit package width support the entire package and allow for entry of the forks of the lift truck.

g. *Stacking lumber in hand-built pile yards.*—Random-width hardwood boards should be spaced. Very wide piles may require flues or a chimney in addition. Random-length hardwood boards will warp less if they are box piled. Chimneys or a number of flues are recommended in building sloped and pitched piles of softwood lumber. Board spacing between flues is usually not essential. Board spacing in addition to a large central chimney is recommended for self-stickered piles of softwood lumber.

h. *Stacking stickers.*—Stickers processed from high-grade dry and surfaced heartwood lumber give longer and satisfactory service. They should be stored under cover when not in use.

3. *Control of Drying Defects*

a. *Chemical brown stain.*—Treating green white pines with enzyme inhibitors is effective. Sawed white pine lumber needs to be stacked promptly and subjected to good air-drying conditions as soon as possible after stacking.

b. *Sticker marking.*—Use dry stickers. The sawed lumber should be stacked and subjected to good air-drying conditions promptly.

c. *Blue stain, mold, and decay.*—Treatment of the green lumber with a suitable fungicide is recommended. Tight pile roofs prevent rain from wetting the lumber. Good yard sanitation practices are essential to keep sources of infection to a minimum.

d. *Insect infestation.*—Logs and lumber may require spray treatment with a suitable insecticide. Insect breeding places need to be cleaned up or destroyed.

e. *End checks and splits.*—Freshly trimmed ends of high value lumber thicker than 5/4 inches should be end coated to minimize defects. Sun shields may be more practical in some situations.

101

f. *Surface checks.*—Presurfacing check-prone species like oak and beech can be considered if grade losses due to surface checking are serious. Package and pile buildup can be modified to retard air movement. Pile spacing can be smaller during periods when the drying potential is severe. Wind baffles may be necessary at times.

g. *Honeycomb.*—Precautions to prevent the development of end checks and surface checks are recommended. Honeycomb in air-dried hardwoods is generally caused by the extension of end checks and the penetration and extension of surface checks.

h. *Warp.*—Good stickering and piling practices are recommended. Uniform-sized stickers, close sticker spacing, and good alinement are essential. Firm and straight foundations with supporting members at all sticker tiers are necessary. Improved thickness control in sawing or presurfacing may be required if miscut lumber significantly contributes to warp development. Clamping or superimposed loading may be a practical remedy for warp-prone species.

4. Cost Reduction

a. *Degrade studies.*—Investigations to evaluate losses in grade and footage caused by drying defects can point out where improvements are needed. Changes in stacking and piling practices may be justified to reduce defects such as surface checking and warp. The treatment of green lumber with a fungicide to control blue stain may be profitable.

b. *Drying time reduction.*—As overall air-drying costs increase with extended yard time, ways and means should be explored to decrease the time that a unit of lumber spends on the yard. A lower inventory being air dried per acre of yard area may be warranted in order to reduce the average drying time.

c. *Labor cost reduction.*—The switch from hand-built pile yarding methods to fork-lift yarding operations results from a need to reduce labor costs. Package handling methods allow increased use of mechanical equipment for stacking, transport, and piling.

5. Protection of Air-Dried Lumber

a. *Outdoor storage.*—If air-dried lumber is at a moisture content of 20 percent or less, it can be bulk piled. However, protection from rain is essential and raintight roofs or tarpaulins can be used. Unit packages of bulk-piled lumber can be wrapped with waterproof paper for temporary storage. If the lumber is at a moisture content more than 20 percent, air drying in place can be continued.

b. *Indoor storage.*—If air-dried lumber is at a moisture content of 20 percent or less, bulk piling indoors in unit packages or in bins is satisfactory. If the lumber is at a high moisture content, the stickered unit packages are stored without dismantling and lumber coming from hand-built piles is stickered. If the shed is not floored, unit package and bin foundations are required to provide ground clearance. Stickering lumber of high moisture content stored in bins would be a good precaution. Circulating air in closed sheds by fans is recommended. Humidistat control is recommended for closed heated sheds.

c. *Truck transit.*—Tarpaulin protection from rain is often used for tractor-trailer transport of air-dried lumber.

d. *Railroad transit.*—Chain cars of bulk-piled strapped packages of air-dried lumber are not protected from the weather. If the lumber was wetted by rains during shipment, prompt stickering for air-drying or kiln-drying is recommended.

GLOSSARY

Active drying period.—A period or season of the year when weather conditions are such that yarded lumber loses moisture rapidly.

Air dried.—The dried condition of lumber, usually 12 to 20 percent moisture content, reached by exposing it for a sufficient period to the prevailing outdoor weather conditions.

Air drying.—Syn: Air seasoning. The process of drying green lumber by exposure to prevailing atmospheric conditions outdoors.

Alleys, cross.—The passageways that connect main alleys in an air-drying yard.

Alleys, main.—Syn: Runways. The roads in an air-drying yard for the transport of lumber.

Alleys, rear.—The space between the backs of rows of hand-built piles in an air-drying yard.

Annual growth ring.—The growth layer put on a tree each year in temperature climates, or each growing season in other climates; each ring includes earlywood and latewood.

Anti-stain chemical.—A chemical applied to lumber to prevent or retard chemical or fungus stain development.

Blue stain.—See: Stain, blue.

Board.—1. Yard lumber that is less than 2 inches thick and 1 or more inches wide. 2. A term usually applied to 1-inch-thick lumber of all widths and lengths.

Board foot.—A unit of measurement equal to a board 1 foot long, 1 foot wide, and 1 inch thick. In finished or surfaced lumber, the board-foot measure is based on the measurement before surfacing or other finishing. In the lumber industry, the working unit is 1,000 board feet; abbr. Mbd. ft.; MBM; MBF.

Bolster.—A square piece of wood usually 4 by 4 inches in cross section, placed between stickered packages of lumber to provide space for the entry and exit of the forks of a lift truck.

Bound water.—Syn: Adsorbed moisture, hygroscopic moisture, imbibed moisture. Moisture that is intimately associated with the finer wood elements of the cell wall.

Bow.—Syn: Camber. A form of warp in which a board deviates from flatness lengthwise but not across the faces.

Boxed heart.—Syn: Boxed pith. When the pith falls entirely within the outer faces of a piece of lumber anywhere in its length, it is said to contain boxed heart.

Bright.—Syn: Unstained. The term is applied to lumber that is free from discolorations. The term bright sapwood is sometimes used to describe sapwood of natural color or in which the stain or discoloration can be removed by surfacing to standard thickness.

Brown stain.—See: Stain, chemical brown.

Bulk pile.—Syn: Solid pile. The stacking of lumber into unit packages, or into bins without vertical spaces or stickers between the layers.

Bunk, carrier.—Specially designed wood beams on which lumber is placed, enabling the straddle truck or carrier to pick up the unit for transport.

Cambium.—The 1-cell-thick layer of tissue between the bark and wood that repeatedly subdivides to form new wood and bark cells.

Canal, resin.—Syn: Resin duct. Intercellular passages that contain and transmit materials. They extend vertically or radially in a tree.

Casehardening.—A condition of stress and set in dry lumber in which the outer fibers are under compressive stress and the inner fibers under tensile stress.

Cell.—A general term for the minute units of wood structure having distinct cell walls and cell cavities including wood fibers, vessel segments, and other elements of diverse structure and function.

Check.—Syn: Drying check, checking. A separation of the wood fibers on any surface of a log, timber, or lumber resulting from tension stresses set up during drying, usually the early stages of drying.

Check, end.—A failure usually in the plane of the wood rays on the end-grain surface of a log, timber, or board resulting from stresses caused by too rapid or excessive end drying.

Check, internal.—See: Honeycombing.

Check, surface.—A check occurring on the tangential surface of a board and extending across the annual growth rings into the interior.

Chemical brown stain.—See: Stain, chemical brown.

Chimney.—See: Flue. A vertical space, more than 6 inches wide, in the center of a hand-built pile extending the length and depth of the pile, intended to facilitate air circulation in the pile.

Collapse.—Syn. Washboarding, crimping. A corrugated appearance of the surface of a piece of wood caused by an irregular drawing together of the cell walls as free water leaves the cavities; occurs only in the early stages of drying green heartwood of low permeability, especially quartersawed stock.

103

Common lumber.—A broad grade of lumber usually including several subgrades, applied to both hardwood and softwood.

Compression failure.—Deformation of the wood resulting from excessive pushing together of the fibers along the grain. It may develop in standing trees due to bending by wind or snow or to internal longitudinal stresses imposed during felling. In surfaced lumber, compression failures appear as fine wrinkles across the face of the piece.

Compression wood.—See: Reaction wood. Abnormal wood formed on the lower side of branches and inclined trunks of softwood trees. As seen on the cross section surfaces of a branch or stem, it appears as relatively wide, eccentric growth rings with little or no demarcation between earlywood and latewood and more than normal amounts of latewood. Compression wood shrinks more longitudinally than normal wood.

Correction, temperature.—An adjustment of the readings of the resistance-type electrical moisture meter to compensate for changes in the temperature of the wood.

Course.—Syn: Layer. A single layer of lumber of the same thickness in a stickered pile or package.

Crib pile.—Stacking lumber to usually form a hollow triangle. The boards lap at the corners.

Crook.—A form of warp in which a board deviates edgewise from a straight line from end to end.

Cross break.—A separation of the wood cells across the grain due to internal strains resulting from the unequal longitudinal shrinkage commonly associated with reaction wood.

Cross section.—See: Section, cross.

Cup.—A form of warp in which there is a deviation from flatness across the width of a board.

Decay.—Syn: Rot, dote. The decomposition of wood substance by **fungi.**

Defects.—Any irregularity or imperfection in a tree, log, bolt, or lumber that reduces its volume or quality or lowers its durability, strength, or utility value. Defects may result from knots and other growth conditions and abnormalities; from insect or fungus attack; from milling, drying, machining, or other processing procedures.

Diagonal grain.—See: Grain, diagonal.

Dipping.—Process of submerging lumber in a dipping vat containing fungicides or other chemicals to prevent stain or decay.

Discoloration.—Change in the color of lumber due to fungus and chemical stains or weathering.

Drying.—Syn: Seasoning. The process of removing moisture from lumber to improve its serviceability in use.

Earlywood.—Syn: Springwood. Wood formed during the early period of annual growth; usually less dense than wood formed later.

Edge piling.—Stacking of wood products on edge, e.g., 2 by 4's, so that the broad face of the item is vertical; usually done to restrain crook.

Edge-to-edge stacking.—Syn: Close piling. In stacking stickered unit packages for air drying, placing the edges of each item in the layer against the adjacent pieces.

Electrodes.—In testing wood for moisture content, devices made of electrically conducting material, usually steel pins, for connecting wood into the electric circuit of an electric moisture meter.

Electrodes, insulated.—In testing wood for moisture content, special electrodes for use with resistance-type electric moisture meters that are coated with an insulating material to limit or control the point of contact between the electrode and the wood.

End coating.—A coating of moisture-resistant material applied to the end-grain surfaces of green lumber to retard end drying and consequent checking and splitting.

End pile.—Stacking of green lumber on end, and inclined, in a long, fairly narrow row, the layers separated by stickers.

End racking.—Placing of boards on end against a support to form a pile in the shape of an "X" or an inverted "V".

End splitting.—See: Split, end.

Equilibrium moisture content.—Abbr: EMC. The moisture content at which wood neither gains nor loses moisture when surrounded by air at a given relative humidity and temperature.

Extractives.—Substances in wood, not an integral part of the cellular structure, that can be removed by hot or cold water, ether, benzene, or other solvents that do not react chemically with wood substance.

Fiber.—A general term for any long narrow cell.

Fiber saturation point.—Abbr: FSP. The stage in the drying of wood at which the cell walls are saturated with water and the cell cavities are free of water. It is usually considered to be approximately 30 percent moisture content, based on the weight of the wood when ovendry.

Flat pile.—Stacking of lumber so that the broad face of the item is horizontal.

Flat pile, sloped.—A pile of stacked lumber on foundations that are not level, but deviate from the horizontal. The slope is usually lengthwise of the board and may be 1 inch or more per foot of length.

Flatsawn.—See: Grain, flat.

Flue.—See: Chimney. Vertical spaces, 6 inches or less in width in the pile and extending the length and depth of the pile, intended to facilitate circulation of air within the pile.

Foundation, open.—Structural supports for the air-drying pile designed to facilitate air circulation under the pile.

Free water.—Syn: Capillary water. Water that is held in the cell cavity of the wood.

Fungi.—Low forms of plants consisting mostly of microscopic threads that traverse wood in all directions dissolving out of the cell wall materials they use for their own growth.

Grade.—A classification or designation of the quality of manufactured pieces of wood or of logs and trees.

Grain.—The direction, size, arrangement, appearance, or quality of the fibers in lumber.

Grain, cross.—Lumber in which the fibers deviate from a line parallel to the sides of the piece. Cross grain may be either diagonal or spiral grain or a combination of the two.

Grain, diagonal.—Lumber in which the annual rings are at an angle with the axis of a piece as a result of sawing at an angle with the bark of the log. A form of cross grain.

Grain, edge.—Syn: Comb grain, edge-sawn, quarter grain, quarter-sawed, rift grain, rift sawed, stripe grain, vertical grain. Lumber that has been sawed or split so that the wide surfaces extend approximately at right angles to the annual growth rings exposing the radial surface. Lumber is considered edge-grained when the rings form an angle of 45° to 90° with the wide surface of the piece.

Grain, end.—The ends of logs or timbers, dimension, and boards that are cut perpendicular to the fiber direction.

Grain, flat.—Syn: Flatsawn, plain grain, plain-sawed, slash grain, tangential cut. Lumber that is sawed or split in a plane approximately perpendicular to the radius of the log. Lumber is considered flat-grained when the annual growth rings make an angle of less than 45° with the surface of the piece.

Grain, spiral.—A form of cross grain in lumber in which the fibers take a spiral course about the trunk of a tree instead of the normal vertical course. The spiral may extend in a right-handed or left-handed direction around the tree trunk.

Grain, straight.—Lumber in which the fibers and other longitudinal elements run parallel to the axis of a piece.

Grained, close.—Lumber with narrow, inconspicuous annual rings. The term is sometimes used to designate wood having small and closely spaced pores, but in this sense the term "fine textured" is more often used.

Grained, coarse.—Lumber with wide, conspicuous annual rings in which there is considerable difference between earlywood and latewood. The term is sometimes used to designate wood with large pores, such as oak, ash, chestnut, and walnut, but in this sense "coarse textured" is more often used.

Hardwoods.—Generally one of the botanical groups of trees that have broad leaves, e.g., oak, elm, basswood. Also, the wood produced from such trees. The term has no reference to the actual hardness of the wood.

Heartwood.—The inner layers of wood which, in the growing tree, have ceased to contain living cells and in which the reserve materials, e.g., starch, have been removed or converted into heartwood substances. It is generally darker in color than sapwood though not always clearly differentiated.

High-frequency dielectric heating.—Heating wood with an electric field oscillating at frequencies of 1 to 30 million cycles per second. The electric energy is applied to the wood between metal plates as electrodes.

High-temperature drying.—In kiln drying wood, use of dry-bulb temperatures of 212° F. or more.

Honeycombing.—Syn: Hollow horning, internal checking, interior checking, inner checking. In lumber, separation of the fibers in the interior of the piece, usually along the wood rays. The failures often are not visible on the surfaces, although they can be extensions of surface and end checks.

Humidity, relative.—The ratio of the vapor pressure of water in a given space compared with the vapor pressure at saturation for the same dry-bulb temperature. Under ordinary temperatures and pressures, it is the ratio of the weight of water vapor in a given space compared with the weight which the same space is capable of containing when fully saturated at the the same temperature.

Hygrometer.—An instrument for measuring the dry-bulb and wet-bulb temperatures of the air, usually consisting of two thermometers.

Infection.—The invasion of lumber by fungi or other micro-organisms.

Infestation.—The establishment of insects or other animals in lumber.

Juvenile wood.—The wide growth, indistinct rings formed near the pith of the tree.

Kiln drying.—The process of drying lumber in a closed chamber in which the temperature and relative humidity of the circulated air can be controlled.

Kink.—A form of warp in which there are sharp deviations from flatness or straightness due to exceptionally abrupt grain distortions, such as around knots, or the piece is sharply bent by misplaced stickers.

Knot.—That portion of a branch or limb which has been surrounded by subsequent growth of the tree.

Latewood.—Syn: Summerwood. The portion of the annual growth ring that is formed after the earlywood formation has ceased. It is usually denser and stronger than earlywood.

Lumber.—The product of the sawmill and planing mill not further manufactured than by sawing, resawing, passing lengthwise through a standard planing machine, cross cutting to length, and matching.

Lumber, nominal size.—As applied to lumber, the rough-sawed commercial size by which it is known and sold in the market.

Meter, electric moisture.—An instrument used for rapid determination of the moisture content in wood by electrical means.

Microwave heating.—Drying a material by using electromagnetic energy alternating at a frequency from 915 megahertz to 22,125 megahertz.

Moisture content.—The amount of water contained in the wood, usually expressed as a percentage of the weight of the ovendry wood.

Moisture content section.—A cross section approximately 1 inch in length along the grain, cut from a board and used to determine moisture content by the ovendry method.

Mold.—A fungus growth on lumber at or near the surface and, therefore, not typically resulting in deep discolorations.

Ovendry.—A term used to describe wood that has been dried in a ventilated oven at $103° \pm 2°$ C. until there is no further loss in weight.

Pile.—Syn: Stack, rack. Stacking lumber layer by layer, separated by stickers or self-stickering, on a supporting foundation (hand stacking) or placing stickered unit packages by lift truck or crane, one above the other on a foundation and separated by bolsters.

Pile pitch.—The forward lean of the front face of a hand-built sloped pile of lumber. The pitch is usually approximately 1 inch in 12 inches of pile height.

Pile roof.—Syn: Pile cover, cover boards, stack cover. A cover on top of the pile to protect the upper layers from exposure to the degrading influences of sun, rain, and snow. The sides and ends of the roof may project beyond the pile to provide added protection.

Pile, box.—A method of flat stacking random-length lumber for air drying. Full-length boards are placed in the outer edges of each layer and shorter boards are alternated lengthwise in between to produce square end piles or unit packages.

Pitch pocket.—Syn: Gum check. An opening extending parallel to the annual growth rings containing, or that has contained, resins, either solid or liquid.

Pith.—The small, soft core occurring in the structural center of a tree trunk, branch, twig, or log.

Rays.—A ribbon like grouping of cells extending radially across the grain, so oriented that the face of the ribbon is exposed as a fleck on the quartersawed surface.

Reaction wood.—Wood with more or less distinctive anatomical characteristics, formed in parts of leaning or crooked stems and branches. In hardwoods, this consists of tension wood and in softwoods compression wood.

Refractory.—Implies difficulty in processing or manufacturing by ordinary methods; resistance to the penetration of preservatives; difficulty in drying or difficulty in working.

Resins.—A class of amorphous vegetable substances secreted by certain plants or trees.

Ring failure.—Syn: Ring shake. A separation of wood along the grain and parallel to the annual rings, either within or between the rings.

Rot.—See: Decay.

Sap.—The moisture in green wood and all that is held in solution.

Sapstain.—See: Stain, blue.

Sapwood.—The outer layers of the stem that in the living tree contain living cells and reserve materials, e.g. starch. The sapwood is generally lighter in color than the heartwood.

Section, cross.—Syn: Transverse section. A section of a board or log taken at right angles to its longitudinal axis.

Shake.—See: Ring failure.

Shrinkage.—The contraction of wood fibers caused by drying below the fiber saturation point. Shrinkage—radial, tangential, and longitudinal—is usually expressed as a percentage of the dimension of the wood when green.

Softwoods.—Generally, one of the botanical groups of trees that, in most cases, have needlelike or scalelike leaves; the conifers; also, the wood produced by such trees. The term has no reference to the actual hardness of the wood.

Solvent seasoning.—A special seasoning process whereby the water in green wood is extracted by a hot or cold solvent that is miscible in water.

Sorting.—Segregation of rough-sawn wood items into lots having similar drying characteristics, such as thickness, species, grades, sapwood, heartwood, grain patterns, and into classes for stacking, such as width and length.

Specific gravity.—The ratio of the ovendry weight of a piece of wood to the weight of an equal volume of water at 4° C. (39° F.). Specific gravity of wood is usually based on the green volume.

Split.—A separation of the fibers along the grain forming a crack or fissure that often extends through the piece from one surface to another.

Split, end.—A split at the end of a log or board. Often an extension of end checks.

Springwood.—See: Earlywood.

Stack.—To make a pile on a suitable foundation by hand-placing layers of lumber on stickers, or similar stock (self-stickered). To construct by hand or with mechanical equipment, unit packages of the item to be dried, by separating layers of the stock with stickers.

Stacking rack.—Syn: Stacking jig, stacking stall. In hand building stickered unit packages of lumber, guides are provided to produce good sticker alinement and square sides.

Stain.—A discoloration in wood that may be caused by micro-organisms, metal, or chemicals.

Stain, blue.—Syn: Sapstain. A bluish or grayish discoloration in the sapwood caused by the growth of certain dark-colored fungi.

Stain, chemical brown.—Syn: Brown oxidation stain, kiln brown stain, yard brown stain, kiln burn. A brownish discoloration of chemical origin in wood that sometimes occurs during the air drying of several softwood species, apparently caused by the concentration and oxidation of extractives.

Sticker.—Syn: Crosser, strip, piling strip, stick. A wood strip placed between courses of lumber in a pile or unit package and at right angles to the long axis of the stock, to permit air to circulate between the layers.

Sticker marking.—Syn: Crosser stain. Indentation or compression of the lumber by the sticker when the superimposed load is too great for the sticker bearing area. Also, sometimes identified as the discoloration caused by blue stain or chemical brown stain in the wood at the location of the sticker.

Summerwood.—See: Latewood.

Surface checking.—See: Check, surface.

Swelling.—Increase in the dimensions of wood due to increased moisture content. Swelling occurs tangentially, radially, and, to a lesser extent, longitudinally.

Temperature, dry-bulb.—Temperature of air as indicated by a standard thermometer.

Temperature, wet-bulb.—The temperature indicated by any temperature-measuring device, the sensitive element of which is covered by a smooth, clean, soft, water-saturated cloth (wet-bulb wick).

Temperature, wet-bulb depression.—The difference in the readings of the wet- and dry-bulb thermometers.

Texture.—The sizes, distribution, and proportional volumes of the cellular elements of which wood is composed; often used interchangeably with grain. Depending on the relative size and distribution of the cellular elements, texture may be coarse (open grain) or fine, even, or uneven.

Tier.—Stacking lumber or other wood products in vertical alinement. Usually refers to sticker alinement.

Tracheid.—See: Vessels. The elongated cells that constitute the greater part of the structure of the softwoods (fibers). Also present in some hardwoods.

Tramways, elevated.—An elevated roadway, often constructed of wood, between rows of piles and well above the pile foundations; to facilitate hand stacking.

Twist.—Syn: Winding, spiral distortion. A form of warp, a distortion caused by the turning or winding of the edges of a board, so that the four corners of any face are no longer in the same plane.

Vacuum drying.—Drying of wood by subjecting it to alternating periods of heated atmosphere or fluid and a reduced pressure, and pressures below atmospheric.

Vapor drying.—Drying wood by subjecting it to hot vapors produced by boiling an organic chemical, such as xylene.

Vessels.—Syn: Tracheid. Tubelike structure of indeterminate length in porous woods; namely, hardwoods.

Wane.—Bark, or the lack of wood from any cause, on any edge of a piece of square-edged lumber.

Warp.—Distortion in lumber causing departure from its original plane, usually developed during drying. Warp includes cup, bow, crook, twist, and kinks, or any combination thereof.

Weathering.—The mechanical or chemical disintegration and discoloration of the surface of lumber that is caused by exposure to light, the action of dust and sand carried by winds, and the alternate shrinking and swelling of the surface fibers with continual variation in moisture content brought by changes in the weather. Weathering does not include decay.

Weight of wood.—The weight of wood depends on its specific gravity and its moisture content. Weight equals the ovendry wood and the moisture it holds. It is expressed as pounds per cubic foot at a certain moisture content or weight per 1,000 board feet at a specified moisture content.

Wetwood.—Syn: Glassy, water core, water soak. Wood with abnormally high water content and with a translucent or water-soaked appearance. This condition develops only in living trees and does not originate through soaking logs or lumber in water.

Wood.—Syn: Xylem. The tissues of the stem, branches, and roots of a woody plant lying between the pith and cambium, serving for water conduction, mechanical strength, and food storage, and characterized by the presence of tracheids or vessels.

INDEX

110